Thermal Physics

Energy and Entropy

Written by distinguished physics educator David Goodstein, this fresh introduction to thermodynamics, statistical mechanics and the study of matter is ideal for undergraduate courses.

The textbook looks at the behavior of thermodynamic variables and examines partial derivatives – the essential language of thermodynamics. It also explores states of matter and the phase transitions between them, the ideal gas equation and the behavior of the atmosphere. The origin and meaning of the laws of thermodynamics are then discussed, together with Carnot engines and refrigerators and the notion of reversibility. Later chapters cover the partition function, the density of states and energy functions, as well as more advanced topics such as the interactions between particles and equations for the states of gases of varying densities.

Favoring intuitive and qualitative descriptions over exhaustive mathematical derivations, the textbook uses numerous problems and worked examples to help readers get to grips with the subject.

DAVID GOODSTEIN is the Frank J. Gilloon Distinguished Teaching and Service Professor Emeritus at the California Institute of Technology. He has extensive research experience in condensed matter physics, and his book *States of Matter* (Prentice Hall, 1975) was hailed as launching this important field. He directed and hosted the popular television series The Mechanical Universe, which was based on his lectures at Caltech.

Thermal Physics
Energy and Entropy

DAVID GOODSTEIN
California Institute of Technology

CAMBRIDGE
UNIVERSITY PRESS

CAMBRIDGE
UNIVERSITY PRESS

University Printing House, Cambridge CB2 8BS, United Kingdom

Cambridge University Press is part of the University of Cambridge.

It furthers the University's mission by disseminating knowledge in the pursuit of education, learning and research at the highest international levels of excellence.

www.cambridge.org
Information on this title: www.cambridge.org/9781107080119

© D. Goodstein 2015

First published 2015

Printed in the United Kingdom by TJ International Ltd. Padstow Cornwall

A catalog record for this publication is available from the British Library

Library of Congress Cataloging in Publication data
Goodstein, David L., 1939– author.
Thermal physics : energy and entropy / by David Goodstein.
pages   cm
Includes index.
ISBN 978-1-107-08011-9 (Hardback)
1. Thermodynamics.   2. Entropy.   3. Statistical mechanics.   I. Title.
QC311.G585 2015
536′.7–dc23
2014021242

ISBN 978-1-107-08011-9 Hardback

Cambridge University Press has no responsibility for the persistence or accuracy of URLs for external or third-party internet websites referred to in this publication, and does not guarantee that any content on such websites is, or will remain, accurate or appropriate.

Contents

Preface

This is a book for students who have some familiarity with general physics and the calculus to learn what thermodynamics and statistical mechanics is all about, starting with the fact that you don't have to know what a system is doing, just how many things it could be doing, to get precise values of its thermodynamic variables, particularly its temperature and pressure. That insight is discussed and explained in the first chapter, along with some simple observations about the peculiar behavior of very large numbers. It uses the perfect gas to illustrate its points.

Chapter 2 looks into the behavior of thermodynamic variables and gives a lesson in partial derivatives, the essential language of thermodynamics. It also considers the various ways of expanding an ideal gas. Chapter 3 deals with the states of matter and the phase transitions between them, and also covers temperature scales, the ideal gas equation of state and the behavior of the atmosphere.

Chapter 4 considers the origin and meaning of the first and second laws of thermodynamics and goes on to discuss Carnot engines and refrigerators and the notion of reversibility. Chapter 5 addresses the probability of finding a system in a particular state and goes on to deal with the partition function and the density of states.

Chapter 6 concerns itself with the various energy functions and includes a neat mnemonic device for remembering them all. Chapter 7 deals with statistical mechanics for fixed and variable N and the grand partition function. Chapter 8 treats some more advanced topics, particularly interactions among the particles and equations of state of gases of varying densities.

The book is designed to offer a good rigorous introduction to thermodynamics, statistical mechanics and the study of matter in general.

Acknowledgments

I am very happy to acknowledge the generous assistance given me by my wife, Dr. Judith Goodstein. It was she who among many other things found all of the photographs in the places given below.

Photographs of James Prescott Joule, James Clerk Maxwell, Rudolf Clausius, Max Planck, Albert Einstein, Michael Faraday, Josiah Willard Gibbs, William Thomson (Lord Kelvin), Nicolas Léonard, Sadi Carnot, Courtesy of the Archives, California Institute of Technology.

Photographs of Ludwig Boltzmann and Johannes Van der Waals, Emilio Segrè, Visual Archives, American Institute of Physics.

Photograph of Joseph and Maria Goeppert Mayer, Mandeville Special Collections, UC San Diego Library.

1

The basic ideas of thermodynamics and statistical mechanics

1.1 From atoms to thermodynamics

Imagine a box, a cube 10 cm on each edge, with 10^{22} atoms of helium gas in it. The atoms share among them some total energy U; say, 2×10^6 ergs, which cannot change because the box is isolated from the rest of the world. Inside the box the atoms fly around, banging into each other or the walls, exchanging energy and momentum. If there is only one atom in the box, and we know how it started out, we might imagine being able to calculate its precise trajectory for a while, predicting just where it would end up at some later time. If there are twenty atoms, the same job becomes horribly more complicated. With 10^{22} atoms it is obviously hopeless. Moreover, according to the laws of quantum mechanics, it would not be possible even in principle. If we knew precisely where the atoms were at some time, we could have no idea of how fast they were moving, according to the uncertainty principle. Obviously, a very short time after we start things off, there is not much we can say about what's going on inside the box.

Nevertheless, it is possible to make some very precise statements about the properties of the gas in the box, especially if we allow some time to pass after we start it off. For example, the gas will have some pressure, P, and some temperature, T, and, given the information we already have, these can be predicted with extreme accuracy and confidence. Temperature and pressure are macroscopic or thermodynamic quantities. The problem before us in this section is to describe the connection between these (predictable) thermodynamic quantities and the (unpredictable) microscopic quantities that somehow give rise to them.

The trick, as it usually is in physics, is to ask the right question. We cannot, even in principle, say exactly what is going on inside the box some time after we isolate it, but we can, in principle at least, say how many possibilities there are. Let us focus our attention on that question.

We have a box of volume V (10^3 cm^3 in our example) containing atoms of some kind (10^{22} atoms of helium) with total energy U (2×10^6 ergs). Let us call the number of things that can possibly happen Γ. That is, there are Γ ways in which N atoms can divide among them energy U while remaining in volume V. (It is not obvious that such a number exists, but it does. We shall see shortly just what we mean by "the number of possible things that can happen".)

If we change U, V or N, the number Γ will change. In other words, Γ is a function of the numbers U, V and N. It will turn out to be convenient to deal not with the (usually gigantic) number Γ but rather with its (more manageable) natural logarithm (written as "log" rather than "ln" throughout this book). We define the quantity S,

$$S = k_B \log \Gamma \qquad (1.1.1)$$

where k_B here, called Boltzmann's constant, will be assigned a value later. The quantity S is called the **entropy**. Since Γ is a function of U, V and N, S is also a function of those variables. If we add more energy to the box, it seems clear that the number of ways of dividing the (larger) energy among the same number of particles must increase. Thus Γ, and hence S, should be a monotonic function of energy at a given V and N. If we knew the functional form we could therefore solve uniquely for U as a function of S, V and N. Let us write

$$U = U(S, V, N) \qquad (1.1.2)$$

We are here supposed to visualize an equation with only U on the left-hand side, and on the right a mathematical form that involves, aside from constants, only the variables S, V and N (not U). Equation (1.1.2) means that any change in U comes about by means of changes of its three variables. Moreover, any small change can be constructed by changing the variables one at a time. We express that fact by writing

$$dU = \left(\frac{\partial U}{\partial S}\right)_{V,N} dS + \left(\frac{\partial U}{\partial V}\right)_{S,N} dV + \left(\frac{\partial U}{\partial N}\right)_{S,V} dN \qquad (1.1.3)$$

The coefficients of dS, dV and dN are called partial derivatives. They are, in effect, defined by this equation. Each partial derivative expresses a precisely defined operation, both physically and mathematically. For example, $(\partial U/\partial S)_{V,N}$ means how much does the energy of the system change if we change the entropy by dS while holding V and N fixed? Mathematically, we are to calculate the derivative of U with respect to S while treating V and N as constants.

Example 1.1.1

Find $(\partial U/\partial S)_{V,N}$ for an ideal gas of atoms.

Solution.

For an ideal gas, Eq. (1.1.2) has the form

$$U = \frac{3}{2}Nk_B \left(\frac{N}{V}\right)^{2/3} \exp\left[\frac{S}{(3/2)Nk_B} - s_0\right] \tag{1.1.4}$$

where k_B is Boltzmann's constant and s_0 is also a constant. So

$$\left(\frac{\partial U}{\partial S}\right)_{V,N} = \left(\frac{N}{V}\right)^{2/3} \exp\left[\frac{S}{(3/2)Nk_B} - s_0\right] = \frac{U}{(3/2)Nk_B}$$

Equilibrium thermodynamics is largely an expression of the fact that the energy of a body is a unique function of S and (generally) one or two other variables such as V and N. The consequences of this fact are in turn expressed by partial derivatives. The mathematics of partial derivatives is the language of equilibrium thermodynamics.

Of the many partial derivatives that will show up in the course of our work, a few have particular significance and are therefore given special names. Among those chosen few are the three coefficients of the differentials in Eq. (1.1.3). We define

$$T = \left(\frac{\partial U}{\partial S}\right)_{V,N} \tag{1.1.5}$$

where T is called the absolute thermodynamic temperature;

$$P = -\left(\frac{\partial U}{\partial V}\right)_{S,N} \tag{1.1.6}$$

where P is called the pressure; and

$$\mu = \left(\frac{\partial U}{\partial N}\right)_{S,V} \tag{1.1.7}$$

where μ is called the chemical potential.

Example 1.1.2

For the system obeying Eq. (1.1.4) find the pressure as a function of T, V and N.

Solution.

From Example (1.1.1) we have

$$T = \left(\frac{\partial U}{\partial S}\right)_{V,N} = \frac{U}{(3/2)Nk_B}$$

and also

$$P = -\left(\frac{\partial U}{\partial V}\right)_{S,N} = +\frac{2U}{3V}$$

(since $U \propto V^{-2/3}$ with everything else held constant). Upon eliminating U between these two equations, we have

$$P = \frac{Nk_B T}{V} \tag{1.1.8}$$

The relation involving P, T, V and N for any system is called the equation of state. Equation (1.1.8) is the equation of state of the ideal gas.

There is a technical point to take care of concerning units. The constant k_B is related to the temperature by

$$T = \frac{\partial U}{\partial S} = \frac{1}{k_B}\frac{\partial U}{\partial \log \Gamma} = \frac{\Gamma}{k_B}\frac{\partial U}{\partial \Gamma}$$

Thus $k_B T$ has the units of energy. The choice of a value for k_B fixes the absolute temperature scale. We shall choose to express T in kelvins (K), which is accomplished by setting

$$k_B = 1.38 \times 10^{-23} \text{ joules/kelvin}$$
$$= 1.38 \times 10^{-16} \text{ ergs/kelvin}$$

On the Kelvin scale, zero is the absolute zero of temperature. Water, ice and water vapor coexist at the unique temperature of 273.15 K, and the normal boiling point of water is exactly 100 K higher. Room temperature (of a rather warm room) is roughly 300 K.

The connection we set out to make has now been made. Starting from a purely microscopic idea – the number Γ of ways that N atoms could divide the available energy, we have shown what is meant by purely macroscopic ideas such as temperature and pressure. We have, of course, not yet shown that the peculiar entities that appear in Eqs. (1.1.5) and (1.1.6) behave as we intuitively feel temperature and pressure ought to behave. That will come shortly. It is unlikely that you have any intuitive feel for chemical potential. We shall try to develop that intuition later.

Problem 1.1
For a system obeying Eq. (1.1.4), find the following functions:

$$S = S(T, V)$$
$$S = S(T, P)$$

Problem 1.2
For an ideal gas of large N and U, obeying Eq. (1.1.4), we wish to carry out the following operation. We add one atom with zero energy so that the gas has the same amount of energy but $N + 1$ atoms. We then wish to extract enough energy that the entropy of the system is the same as it was before the atom was added. How much energy must be extracted?

1.2 Counting quantum states

We have seen that, in order to connect thermodynamics to the microscopic world of atoms and molecules, the question we must answer is not what the atoms are doing, but rather, how many things can they be doing? In this section, we shall see exactly what is meant by that question and by its answer, the number Γ, in the case of the perfect gas.

The perfect gas is one whose atoms exert no forces on one another. It is a good approximation to the behavior of real matter at low densities and high temperatures. In those conditions it becomes the same as the ideal gas of Example 1.1.1. Our interest in it now, however, is as a model, an idealization that will help us form more concrete ideas about how to describe the microscopic behavior of matter.

To begin with, we consider the simpler problem of a single atom confined in an otherwise empty box in the form of a cube whose dimension is L on each edge. The energy of the atom is simply its kinetic energy, which is related to its momentum by

$$\varepsilon = \frac{p^2}{2m} \tag{1.2.1}$$

where ε is the energy, m the mass and $\vec{p}$ the momentum is a vector with x, y and z components, so that

$$p^2 = p_x^2 + p_y^2 + p_z^2 \qquad (1.2.2)$$

In classical mechanics each component of $\vec{p}$ is a continuous variable that can take on any positive or negative value. Thus even for the simple problem of a single atom with fixed energy, there would be no answer to the question of how many ways the atom could use up the energy it has. For any finite ε there are an infinite (i.e. uncountable) number of choices of p_x, p_y and p_z. In quantum mechanics, however, the components of the momentum are quantized and are given in our case by

$$\begin{aligned} p_x &= n_x p_0 \\ p_y &= n_y p_0 \\ p_z &= n_z p_0 \end{aligned} \qquad (1.2.3)$$

where n_x, n_y and n_z are numbers (called **quantum numbers**), and p_0 is the quantum unit of momentum in our cubical box. The permissible values of n_x, n_y and n_z and the size of p_0 depend on how we choose to describe the walls of the box.

One way to describe the walls is simply to say they are impenetrable. Thus an atom hitting one of these walls bounces off, conserving energy. An impenetrable wall is sketched in Fig. 1.1. With this specification, each of the ns can be any positive integer,

$$n_x, n_y, n_z = 1, 2, 3, \ldots \text{ (impenetrable walls)}$$

and p_0 is given by

$$p_0 = h/(2L) \text{ (impenetrable walls)}$$

where h is Planck's constant,

$$h = 6.62517 \times 10^{-27} \text{ erg s}$$

(We often use the symbol $\hbar = h/(2\pi) = 1.05 \times 10^{-27}$ erg s.)

All of the possible states of the atom in our box can be enumerated by assigning positive integers to n_x, n_y and n_z. Although our description of the problem is simple and straightforward, this set of solutions has some bizarre aspects, even aside from the fact that ε and $\vec{p}$ are quantized. For one thing, it is impossible for the atom to have zero kinetic energy. The lowest value that the energy can have, say ε_m, occurs when $n_x = n_y = n_z = 1$, so that

$$\varepsilon_m = 3\frac{(h/(2L))^2}{2m} = \frac{3}{8}\frac{h^2}{mL^2}$$

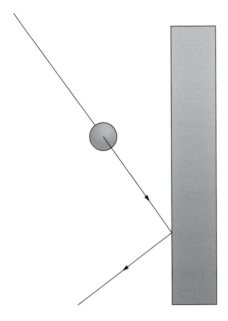

Figure 1.1 An impenetrable wall.

This quantity is called the **zero-point energy**, and it always occurs in quantum mechanics when a particle is confined in space. Even more peculiar, the components of momentum have only positive values. It's hard to see how we can make use of this description to discuss an atom that can fly either to the left or the right.

There is another way of describing the walls that denotes states that do not have these strange quirks. In this description, when the atom hits the wall, it does not bounce off of it at all. Instead it vanishes into the wall, reappearing with the same energy and momentum at the opposite wall, as sketched in Fig. 1.2. In other words it behaves as if the left-hand end of the box always begins just where the right-hand end stops, and there are no walls at all. A problem described this way is said to have **periodic boundary conditions**.

Strange though such a box may seem, it has all the necessary properties for our purposes. Imagine it to have N atoms in it with total energy U. That makes it suitable for thermodynamic analysis. For periodic boundary conditions, the unit of momentum is

$$p_0 = h/L \text{ (periodic boundary conditions)} \qquad (1.2.4)$$

and the allowed values of quantum numbers are

$$n_x, n_y, n_z = 0, \pm 1, \pm 2, \ldots \text{ (periodic boundary conditions)} \qquad (1.2.5)$$

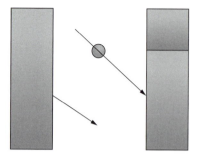

Figure 1.2 Periodic boundary conditions.

We thus have, in this case, a state of zero energy, and both positive and negative components of momentum. Because of these nice properties, we will always use periodic boundary conditions. We are now in a position to see by example, at least for the simple case of one atom in a box, what exactly is meant by the number Γ.

Example 1.2.1

Find Γ for one atom in the box we have been discussing, if the atom has 3 units of energy.

Solution.

The quantum unit of energy in the box is

$$\varepsilon_0 = p^2/(2m) = h^2/(2mL^2) \tag{1.2.6}$$

Using Eqs. (1.2.1), (1.2.2) and (1.2.3), the possible energies of an atom in the box may be written as

$$\varepsilon = \varepsilon_0\left(n_x^2 + n_y^2 + n_z^2\right) \tag{1.2.7}$$

The statement that the atom has 3 units of energy means

$$n_x^2 + n_y^2 + n_z^2 = 3$$

Γ is the number of choices of (n_x, n_y, n_z) that satisfy this last equation. The choices that work are all possible combinations of $n_x = \pm 1$, $n_y = \pm 1$ and $n_z = \pm 1$. There are $2^3 = 8$ possible sets that work, so in this simple case $\Gamma = 8$.

Problem 1.3

One atom in the same box has energy $\varepsilon = B\varepsilon_0$. Find Γ if $B = 0$, and if $B = 1$ or 2 or 4. Find Γ if $B = 25$.

We do not need thermodynamics to discuss the behavior of one atom in a box. But we do need it if there are many atoms. The problem we want to analyze, the perfect gas, is formulated as follows. In the box we have been describing there are N atoms. The possible states of each atom, however, are those it would have if it were alone in the box. In other words, each atom has kinetic energy only. The kinetic energy of each atom is quantized, and its possible values are given by Eq. (1.2.7) with n_x, n_y and n_z each equal to zero or any positive or negative integer. Thus the quantum mechanical description of the problem is no different from what it was before, but the problem of counting how many ways a given amount of energy can be allocated has become dramatically more difficult.

Before going on with this discussion, we must come to grips with a purely linguistic difficulty. The problem is that the word *state* simply has too many uses. We speak of the state of a single atom, the microscopic state of a gas of atoms, the macroscopic state of a gas (i.e. its temperature and pressure), the liquid state, the solid state; and, although Philadelphia is in the Commonwealth of Pennsylvania, San Francisco is in the state of California – and you may now be in a state of confusion. In the hope of denting that confusion a little bit, we will now replace one of those uses of the word with a special term for our purposes. We will refer to a microscopic state of a single particle as a *level*.

A level is a particular set of quantum numbers. Thus $n_x = 2$, $n_y = -3$, $n_z = 0$ is a level, which we can call "the level $(2, -3, 0)$". If a particle is in that state, we will say it "occupies the level $(2, -3, 0)$". Any atom occupying that level has energy $\varepsilon = 13\varepsilon_0 = 13h^2/(2mL^2)$. There may be many levels with the same energy for a single particle. In fact, the last part of Problem 1.3 can be restated as follows: How many levels are there with energy $25\varepsilon_0$ for one particle?

If there is more than one atom (or molecule or particle) in the box, there may be more than one atom in the same level. We will call the number of atoms in a level the *occupation number* of the level. Giving the occupation numbers of all of the levels specifies the microscopic state of a system of particles in a box. In other words, we need to know how many particles are in each level, but not which ones are there.

Example 1.2.2

Here is a schematic representation of one possible microscopic state of a system consisting of six atoms with a total of 7 units of energy:

n_x	n_y	n_z	Energy of each atom in this level	Occupation number	Total energy of atoms in level
0	0	0	0	2	0
1	0	0	ε_0	0	0
0	1	0	ε_0	1	ε_0
0	0	1	ε_0	0	0
−1	0	0	ε_0	0	0
0	−1	0	ε_0	0	0
0	0	−1	ε_0	0	0
...	...	...			
1	1	0	$2\varepsilon_0$	3	$6\varepsilon_0$
			All other states Unoccupied		

We are now in a position to say exactly what is meant by Γ for a perfect gas in the situation which formed our starting point: N atoms in an isolated box with total energy U. For convenience we take the box to be a cube of side L, so that $V = L^3$, and $U = B\varepsilon_0$, where B is some (usually very big) number (in Example 1.2.2 above, $N = 6$ and $B = 7$). A possible microscopic state of the system is a specific set of all the occupation numbers of all the levels in the box such that all the particles get used up, and all the energy gets used up. Γ is then the number of possible microscopic states of the system.

Problem 1.4

For the situation outlined in the paragraph above, find Γ in the following cases:

(a) $N = 2$ and $B = 24$ (answer   4116)

(b) $N = 2$ and $B = 25$ (answer   3906)

(c) $N = 2$ and $B = 26$ (answer   5040)

Hint: The problem is just to organize the job of counting. Consider, for example, the case in which $B = 25$. In every possible state of the system, two levels are occupied by one atom each (since the energies must add up to an odd number) and all other levels are empty. First consider all states in which the level $(0, 0, 0)$ is occupied. The other atom must be in some level with 25 units of energy. The possibilities are $(\pm 5, 0, 0)$, $(0, \pm 5, 0)$, $(0, 0, \pm 5)$; six levels so far. We can also reach 25 with a combination of 3^2 and 4^2. In the vector (n_x, n_y, n_z), there are three places for a 4 to appear, and for each of those, two remaining places for the 3. That makes six arrangements. For each of those, each of the nonzero quantum numbers can be positive or negative, $2^2 = 4$ possibilities for each of the six arrangements, a total of $4 \times 6 = 24$ levels in all. Thus the total number of levels in which one particle is in level zero is $6 + 24 = 30$. Next, count up the number of states in which one particle is in one of the states that has one unit of energy, the other on some level with 24 units ... and so on.

The purpose of doing the exercise above is to obtain a firm grasp of the meaning of Γ, the central concept connecting the microscopic world to thermodynamics. In any problem of real interest, B will not be 25, but much bigger (B stands for big) and Γ will be unimaginably large (Γ stands for gigantic). Counting states as we have done is never a practical way of doing real problems.

Even though we never actually count states by hand as we have done here, the idea that there exists a definite number of possible states in an isolated system is the central concept of statistical mechanics. The scientist who took the most important first step toward trying to formalize that point of view was Ludwig Boltzmann (1844–1906). He spent much of his life trying to master the behavior of gases. When he died, by suicide, he

Ludwig Boltzmann

Max Planck

asked that his epitaph be the equation $S = k_B \log W$, which he considered his principal contribution to human knowledge. Since Boltzmann's description was necessarily in terms of classical mechanics, his W was not quite the same as our Γ. Nevertheless, it has entered into the folklore of physics that our fundamental equation Eq. (1.1.1) was given to us by Boltzmann, then rendered meaningful by Max Planck (1858–1947), who was the first to show that energy was quantized. But in point of historical fact, it was Boltzmann, not Planck, who first imagined energy to be quantized, and it was Planck, not Boltzmann who first used Eq. (1.1.1) as we now understand it. These are by no means the only curiosities in the complex and involuted history of this subject, into which we shall dip from time to time.

Even after Boltzmann and Planck had done their work, one more fundamental discovery, perhaps the most subtle of all, remained to be made before the counting of states could be formulated as we have done. The point, which was first grasped by Josiah Willard Gibbs (1839–1903) but put firmly in place by Albert Einstein (1879–1955), is that atoms of the same kind are indistinguishable from one another. The indistinguishability of atoms alters the ways in which states can be counted and the number of states available.

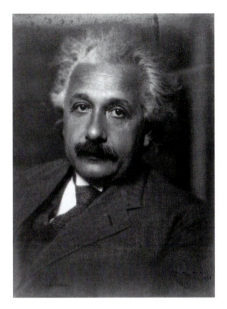

Albert Einstein

Example 1.2.3

Suppose that in Problem 1.4, part (b), the two particles that are to divide between them 25 units of energy are somehow distinguishable from each other. How many states would then be possible?

Answer: Twice as many as before, i.e. $2 \times 3906 = 7812$. In the previous (correct) way of counting states, there was, for example, one state of the system in which the levels $(0, 0, 0)$ and $(5, 0, 0)$ were occupied. If the atoms were distinguishable so that we could call them, say, atom a and atom b, we would have one state with atom a in $(0, 0, 0)$ and atom b in $(5, 0, 0)$, and another with atom a in $(5, 0, 0)$ and atom b in $(0, 0, 0)$. In this way, for each state of indistinguishable atoms we get two states if they are distinguishable.

Problem 1.5
How many states would we get if the atoms were distinguishable in parts (a) and (c) of Problem 1.4? Hint: The answer is less than twice as many in each case.

The important difference between the two ways of counting states is this: The correct method for indistinguishable atoms is to specify only *how many* atoms occupy each level. The incorrect method takes account of *which* atoms are in each level.

In this section we have introduced the *perfect gas*. Later we will discover that, at low density and high temperature, any gas obeys the equations of the *ideal gas* given in Example (1.1.1). The ideal gas is identified by its thermodynamic behavior, for example by the equation $PV = Nk_BT$, which applies also to the perfect gas. Atomic and molecular ideal gases differ in other ways, but both obey this equation of state. The perfect gas is a quantum mechanical model, which does not always behave like an ideal gas. We shall retain this distinction between perfect and ideal gases throughout this book.

1.3 Equilibrium, equal probabilities and temperature

Now that we understand the meaning of the number Γ, we are prepared to see why the quantity $T = (\partial U / \partial S)_{V,N}$ may sensibly be called a temperature. This is a crucial step in the connection between microscopic physics and thermodynamics. To get right to the heart of the matter, let us postpone temporarily a discussion of temperature scales, means of measuring temperature and the like. If we place in contact two bodies that are already at the same temperature, there will be no net flow of energy between them. Otherwise, energy flows from the hotter to the cooler. The question at hand is why the peculiar quantity $(\partial U / \partial S)_{V,N}$ tells us whether energy will flow and, if so, in which direction.

To make this connection, we must introduce two fundamental assumptions, upon which rests the entire edifice of statistical mechanics and thermodynamics. These are the two assumptions.

1. An isolated system (such as a box of gas) will eventually reach **equilibrium**.
2. Once in equilibrium *all permissible quantum states of the system are equally probable*.

Here is what is meant by the two assumptions. Suppose we take our box of gas, add energy to it by, say, heating it from below, and then isolate it so that no energy can get in or out. At first the energy will be concentrated in the atoms near the bottom of the box, but, after a while, it will spread out so that an observer would have no way of telling whether the gas got its energy by being heated at the bottom rather than at the top or on one side. This condition, wherein the body has lost all memory of its past, is what we mean by *equilibrium*. At that point we can view the system microscopically and pose a purely quantum mechanical problem: What are the physically permissible states of the system? For example, in the perfect gas of the previous section, any state is physically permissible if it uses up all of the energy U in the system by distributing it among the N atoms constrained to remain in the volume V. As we saw, there is some number of such states, which we call Γ. The second assumption is that each of these states has the same probability.

We know intuitively that if a reasonable box of gas has N atoms and energy U, and we look at it at some instant, we can expect to find that the atoms have shared the energy out among them. That is, each atom can be expected to have roughly the average energy per atom, $\bar{\varepsilon} = U/N$. Our assumption, however, is that a state in which that is true has exactly the same probability as one in which all the atoms except one are lying inert on the bottom of the box, and that one active atom is buzzing madly about, carrying all the energy in the system. Would it not be better to construct our edifice on a more reasonable foundation?

The edifice stands, and is in fact the most precise and successful of all the sciences. The reason is that, for a macroscopic system, the number of states in which the energy is spread out sensibly is so large compared with the number of perverse and unexpected states that the latter are simply never observed. The point is a purely quantitative one, to which we shall have to return later for verification, but what it rests on is this: For any macroscopic system, Γ is a number that is large beyond comprehension.

Leaving aside for now this quantitative question, let us see why, under our two assumptions, $(\partial U/\partial S)_{N,V}$ has the properties of a temperature. Imagine two separate bodies. Each of them could, for example, be a box of perfect gas. Label the bodies 1 and 2. As indicated in Fig. 1.3, they have respectively N_1 and N_2 atoms in volumes V_1 and V_2 with energies U_1 and U_2. As a consequence, their numbers of possible states are Γ_1 and Γ_2,

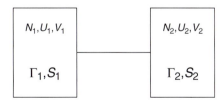

Figure 1.3 Two boxes of gas connected by a wire.

giving rise to entropies S_1 and S_2. They are isolated from the rest of the world, but can be connected to each other by means of, say, a copper wire. When they are connected, energy can flow through the wire from one body to the other, but the total energy of the system, $U = U_1 + U_2$, never changes. Moreover, N_1 and N_2, V_1 and V_2 never change. The question is: What is the condition necessary to ensure that, when we connect the two boxes, no energy will flow through the wire?

Let the two boxes be disconnected, each one internally in equilibrium. Then how many states can the system as a whole have? The answer is

$$\Gamma = \Gamma_1 \Gamma_2 \qquad (1.3.1)$$

So, the entropy of the combined system is

$$S = k_B \log \Gamma$$
$$= k_B \log \Gamma_1 + k_B \log \Gamma_2 = S_1 + S_2 \qquad (1.3.2)$$

Now we momentarily reconnect the two boxes, disconnect them again, and ask: Did any energy flow while they were connected? There is no physical basis for predicting what the result of the experiment will be, but we can say what the most probable result is.

The quantity that is free to change while the two boxes are in contact is the fraction of the energy in each one. Let us call the fraction of the energy in box 1 x. Thus,

$$U_1 = xU$$
$$U_2 = (1 - x)U \qquad (1.3.3)$$

The number of possible states in each box depends on x:

$$\Gamma_1 = \Gamma_1(x)$$
$$\Gamma_2 = \Gamma_2(x)$$

so that the number of states in the combined system also depends on x:

$$\Gamma = \Gamma(x)$$

While the boxes are connected, if energy flows from one to the other, x changes. Since by our assumption all permissible states of the combined system are equally likely, the most probable value of x when we disconnect the two boxes is the one for which Γ has the largest value. Therefore, the condition that no energy flows when we disconnect the two boxes is just that x have the value for which Γ is a maximum.

The value of x that maximizes Γ will also maximize $S = k_B \log \Gamma$. Thus the condition we seek is $dS/dx = 0$, or, using Eq. (1.3.2),

$$\frac{dS_1}{dx} + \frac{dS_2}{dx} = 0 \qquad (1.3.4)$$

Any change in x causes changes in U_1, S_1, U_2 and S_2. These are related by

$$\frac{dU_1}{dx} = \left(\frac{\partial U_1}{\partial S_1}\right)_{N_1, V_1} \frac{dS_1}{dx} \qquad (1.3.5)$$

$$\frac{dU_2}{dx} = \left(\frac{\partial U_2}{\partial S_2}\right)_{N_1, V_1} \frac{dS_2}{dx} \qquad (1.3.6)$$

But, since $dU_1 + dU_2 = 0$, we have,

$$(\partial U_1/\partial S_1)_{N_1, V_1}\, dS_1/dx + (\partial U_2/\partial S_2)_{N_2, V_2}\, dS_2/dx = 0 \qquad (1.3.7)$$

Substituting Eq. (1.3.4) into Eq. (1.3.7) gives

$$\left[\left(\frac{\partial U_1}{\partial S_1}\right)_{N_1, V_1} - \left(\frac{\partial U_2}{\partial S_2}\right)_{N_2, V_2}\right] \frac{dS_1}{dx} = 0 \qquad (1.3.8)$$

But dS_1/dx is never zero. If x increases, the energy in body 1 increases, there must be more ways to divide it, so Γ_1 increases and thus S_1 increases. Consequently the quantity in brackets must be zero. Our definition of temperature for either box is

$$T = \left(\frac{\partial U}{\partial S}\right)_{N, V}$$

Therefore, the value of x for which no further energy flow is to be expected is just that for which

$$T_1 = T_2$$

with each temperature given by our definition.

1.4 The way entropy works

The little argument of the last section really forms the intellectual core of this book. We shall refer back to it many times or repeat it with variations. It contains within it the essential ideas that underlie thermo-dynamics. Let us now examine those ideas.

At the outset of this chapter, we abandoned the kind of physics that tries to predict the microscopic state of a system, and formulated instead a new physics that is content to know only the number of possible microscopic states. At that point we have already incorporated the idea of thermodynamic equilibrium. The number of possible states Γ is of no interest unless they are all possible. The number Γ cannot be important if the system just has not had time to get into some of them. Therefore, quantities such as temperature and entropy, whose very definitions depend on Γ, apparently have no meaning except in equilibrium.

The argument of the last section, however, suggests a way of discussing a system that is not in equilibrium. There we had a system consisting of two parts, each internally in equilibrium, thus having a well-defined T and S, *but out of equilibrium with each other*. By a simple extension of the argument, a body can be divided mentally into any number of parts, each internally in equilibrium, but not necessarily in equilibrium with one another. Take, for example, a rod of material hotter at one end than at the other. We can mentally chop it into a series of slices, as sketched in Fig. 1.4. Each slice is internally in equilibrium with a definite temperature (and entropy) with an imaginary thermal link (like the copper wire of the last section) to the slice next to it. The real situation in the rod can be approximated as nearly as we like by making more and more thinner and thinner slices.

Thought of in this way, we can see that the possible number of states of a system has some meaning, even when the system is not in equilibrium. Suppose we number the slices of our rod 1, 2, …, n counting from the left (where the temperature is T_L) to the right (where the temperature is T_R). The first slice, at T_L, has some number of states, Γ_1. The second, at $T_L - \delta T$, has Γ_2, and so on up to the nth slice at T_R with Γ_n possible states. Then, obviously, $\Gamma = \Gamma_1\Gamma_2\ldots\Gamma_n$. The entropy of the system is

$$S = k_B \log \Gamma = S_1 + S_2 + \cdots + S_N \qquad (1.4.1)$$

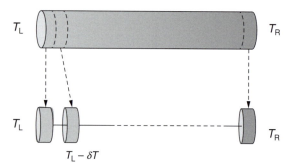

Figure 1.4 A mentally sliced-up rod.

Irrespective of whether the system is in equilibrium or not, the entropy is the sum of the entropies of its parts.

Now, in the situation we have depicted, with the temperature higher toward the left end of the rod, energy will flow to the right from slice to slice. The reason is exactly the same as in the previous section. As energy flows out of slice 1 into slice 2, Γ_1 decreases, Γ_2 increases and the product $\Gamma_1\Gamma_2$ increases. Energy flows from slice 2 to slice 3 once again because that leads to more choices and is therefore more probable, and so on. Energy continues to flow along the rod (if it is isolated from the rest of the world) until $\Gamma = \Gamma_1\Gamma_2\ldots\Gamma_n$ is as big as it can be. That condition is met when the temperature is uniform throughout the rod. The rod is then in its most probable state and no further changes will occur. That state is called thermodynamic equilibrium. The whole process can be described by saying that the entropy of the isolated system, $S = k_B \log \Gamma$, increases until it reaches its maximum value, and then becomes constant. The statement that the entropy of an isolated system can only increase or remain constant is considered by some people to be the most fundamental of all the laws of physics. It is called the second law of thermodynamics.

Let us consider an example of how the principle of maximum entropy may be used.

Example 1.4.1
Consider a box divided into two parts by a moveable partition as shown in Fig. 1.5. Suppose there is gas on both sides of the partition. What determines its equilibrium position?

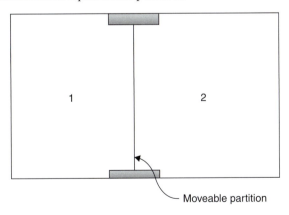

Figure 1.5 A box of gas with a moveable partition.

Solution.
We know that the answer is that the pressure on the two sides must be equal. Let's use the second law to show that the quantity we've defined as $P = -(\partial U/\partial V)_{S,N}$ is the same on both sides.

When the partition moves, no particles change sides, but the energies, entropies and volumes of the two sides all change. We have

$$dU_1 = T\,dS_1 - P_1\,dV_1$$
$$dU_2 = T\,dS_2 - P_2\,dV_2$$

We have taken T to be the same on both sides of the partition because we know that is required for equilibrium. Also, $dN_1 = dN_2 = 0$, and each of the quantities P_1 and P_2 is defined to be $-(\partial U/\partial V)_{S,N}$ for that part of the box. Add together the two equations, recalling that the overall energy and volume are conserved, and we get

$$T\,dS = (P_2 - P_1)dV_1$$

where $dS = dS_1 + dS_2$. We can describe the motion of the partition by the volume it leaves to its left, V_1. Our job, then, is to maximize the entropy with respect to V_1:

$$\frac{\partial S}{\partial V_1} = \frac{P_2 - P_1}{T} = 0$$

The general condition for equilibrium is therefore $P_1 = P_2$, where P is defined by our partial derivative.

Problem 1.6
Suppose that the partition in the above example cannot move, but has a hole in it. Show that in equilibrium the chemical potential is the same on both sides of the box.

Problem 1.7
In mechanics, equilibrium means resistance to change. For example, a marble in a bowl is in stable equilibrium, but a marble balanced on top of an inverted bowl is not (it may be in equilibrium, but it is unstable). Explain the relationship between this kind of equilibrium and the kind we have been discussing here.

There is a point that needs to be emphasized because, while it appears easy to understand now, it becomes more complicated later on. The entropy of an object tends to a maximum *only if the object is isolated.* We shall later see examples in which an object spontaneously changes into states of lower entropy – for example a gas condenses into a liquid, or a liquid freezes into a solid. Such changes occur not in defiance of the principle of maximum entropy, but because of it. They occur when a decrease in the entropy of a part of a system increases the entropy of the whole. We have already seen examples of that phenomenon. In order to increase the entropy of the rod in Fig. 1.4, energy flows from left to right. While that happens the entropy of the first slice on the left, $S = k_B \log \Gamma_1$ is actually decreasing with time. The first slice is not therefore violating the second law of thermodynamics. It is not isolated, but connected to a larger system. As energy flows out of the slice the increase in the total entropy, S, is larger than the decrease in S_1. The first slice unselfishly sacrifices its own entropy for the common good. It is the same principle that causes gases to condense and liquids to freeze.

1.5 Some difficulties and their resolutions

Before we construct the science of thermodynamics on the arguments of this chapter, let us consider three fundamental difficulties that we have so far avoided mentioning.

1. We can see from our discussion that the law of increase of entropy really only describes how a system will *probably* behave. Can it really be true that the most fundamental law of physics is only probably true, and must therefore sometimes be false?
2. In our previous section, our assertion that $\Gamma = \Gamma_1 \Gamma_2$ was strictly true because the states of the two boxes were *independent*. We guaranteed that in turn by the device of the thermal link (the copper wire) that could be disconnected. For our two boxes, when they are connected, or for our imaginary slices of the rod (which are always connected), the total number of possible states is not strictly given by the product of the states, nor is the entropy the sum of the separate entropies.
3. The idea we started with, namely that an isolated system could have a definite energy, is itself an idealization, which can never be strictly correct. Quite aside from the practical difficulty of perfectly isolating anything, the energy must always have some uncertainty, for two reasons. One is thermodynamic: The energy must have been placed in the system by means of a process subject to the probabilistic laws we have been discussing, and therefore only has a most probable value. But, even if we can somehow get around that, there is a quantum mechanical reason why the energy cannot have an exact value.

Quantum mechanically, a system with an exact energy is said to be in a *stationary state*. That means it is trapped forever in a single microscopic state. It has no means of getting into any of the other $\Gamma - 1$ states available at the same energy. Its Γ possible states are therefore not equally probable, and equilibrium can never be attained. This is a result of the uncertainty principle in the form

$$\delta E \, \delta t \geq \hbar \qquad\qquad (1.5.1)$$

where δE is the uncertainty in the energy of the system and δt is the time it spends in one microscopic state. If $\delta E = 0$, $\delta t = \infty$ and the system never changes its state.

In practice this is not a serious problem. For any macroscopic system, an uncertainty in the energy much smaller than the one already present for thermodynamic reasons is quite enough to satisfy Eq. (1.5.1) with δt small enough to let it fluctuate among all its permitted states as required by our assumption.

In spite of all these quibbles, it is nevertheless true that

(1) the second law of thermodynamics works exceedingly well and deserves its place in the shrines of physics,
(2) the entropy of a system is quite accurately given by the sum of the entropies of its parts, and
(3) any deduction drawn from the idea of a definite energy leading to a definite number of states in a large system will be accurately correct.

The basic reason why all three of these principles can be ignored (in, we always emphasize, macroscopic systems) is the mind-boggling immensity of the number Γ. That number is so big that it behaves in ways that we are unaccustomed to. What we mean is this: In a typical system Γ might be a number like 10^N, where N is the number of atoms, say, $N \approx 10^{24}$. Now suppose we have made a mistake in calculating Γ because we have ignored the second or third of the objections raised above. Suppose, moreover, that the error is a very serious one, that in fact we have gotten Γ wrong by a factor of 10^{20} (nobody's perfect). What we're really interested in is not Γ, but rather the entropy, S. We have mistakenly come to the conclusion that

$$S = k_B \log \Gamma = k_B \log 10^{10^{24}}$$
$$\approx 2.3 \times 10^{24} k_B$$

whereas the right answer had we not made our spectacular goof should have been

$$S = k_B \log(10^{10^{24}} \times 10^{20}) = k_B \log 10^{(10^{24} + 20)}$$

$$= (2.3 \times 10^{24} + 20)k_B$$

The error in S thus appears roughly in the 23rd decimal place. For most purposes, this would not be considered a serious error.

The numbers we have used here are pure invention. The real point can only be made by taking real numerical examples. First we consider an example that illustrates the difficulties of exact energies. We'll show how we can smooth it over without changing any important part of our scheme.

In Problem 1.4 we considered a cubic box with two atoms and 24, 25 and 26 quantum units of energy. For those cases let us find the entropy and the temperature. In this problem we are dealing with small values of Γ and exact energies. As noted above, we can expect peculiar results.

The entropies are, respectively (using the solutions given with the problem), for $B = 24$, $\Gamma = 4116$, $S = k_B \log 4116 = 8.33k_B$, for $B = 25$, $\Gamma = 3906$, $S = k_B \log 3906 = 8.27 k_B$, and for $B = 26$, $\Gamma = 5112$, $S = k_B \log 5112 = 8.53k_B$. To find the temperature we must evaluate a derivative,

$$T = \left(\frac{\partial U}{\partial S}\right)_{N,V} = \frac{\Gamma}{k_B} \frac{\partial U}{\partial \Gamma}$$

which can best be estimated by letting U change by one unit at a time. Once again calling the unit of energy ε_0, we have for the step from $B = 24$ to $B = 25$

$$T = -\frac{4011}{210} \frac{\varepsilon_0}{k_B} = -19.1 \frac{\varepsilon_0}{k_B}$$

where we have used 4011, the mean value of the two Γs, and, for the step from $B = 25$ to $B = 26$,

$$T = \frac{4509}{1206} \frac{\varepsilon_0}{k_B} = 3.74 \frac{\varepsilon_0}{k_B}$$

Not only are these two results different from each other, the first is actually negative. This is because Γ declined slightly as the number of units went from 24 to 25. At the beginning of this chapter we blithely assumed Γ would always increase when we increased the energy. The difficulty here is due not to the small Γs, but rather to the exactness of the energy. Suppose we modify our descriptions slightly and say the system must have one unit of uncertainty in the energy. What do we find for the entropy and temperature?

First consider the case $B = 24$ or 25. Then $\Gamma = 4116 + 3906 = 8022$,

$$S = k_B \log 8022 = 9.00 k_B$$

For the case $B = 25$ or 26, $\Gamma = 3906 + 5112 = 9018$,

$$S = k_B \log 9018 = 9.12 k_B$$

On comparing these with the previous results, where U was exact, we see that Γ has been essentially doubled by introducing the uncertainty, but S has changed only by about 10 percent. Even for these very small numbers S is not very sensitive to how precisely we specify U.

The temperature is now given by $\overline{\Gamma} = 8520$ and $\Delta\Gamma = 996$, so

$$T = \frac{8520}{996} \frac{\varepsilon_0}{k_B} = 8.55 \frac{\varepsilon_0}{k_B}$$

In Example 1.1.2 we considered the properties of an ideal gas. An ideal gas is essentially the system we are looking at here, except that there are, say, 10^{23} atoms instead of 2, and an even larger number of units of energy instead of 25. In that limit we found

$$T = \frac{U}{(3/2)N k_B}$$

If, the small number of atoms notwithstanding, we use this formula to estimate the temperature of our box of two atoms and (about) 25 units of energy, we get

$$T = \frac{25}{3} \frac{\varepsilon_0}{k_B} = 8.3 \frac{\varepsilon_0}{k_B}$$

compared with the calculated result $T = 8.55\, \varepsilon_0/k_B$. Thus even the temperature differs by only a few percent. The principal difficulty in applying thermodynamics to Problem 1.4 is not the smallness of the numbers, but rather the exactness of the energy.

Now let us look at the size of the numbers that occur for a realistic ideal gas. To do so we will have to make use of formulas that we will derive in detail only later. For an ideal gas for which N is large and U/N is much larger than ε_0, we can use Eq. (1.1.4). On solving for the entropy, we have

$$S = \frac{3}{2} N k_B \log \left[\frac{U}{(3/2)N k_B} \left(\frac{V}{N} \right)^{2/3} + \frac{3}{2} N s_0 \right]$$

We will find later that

$$\frac{3}{2} N s_0 = N k_B \left[\frac{5}{2} + \frac{3}{2} \log \left(\frac{m k_B}{2 \pi \hbar^2} \right) \right]$$

On putting these together with the result from Example 1.1.2 that $T = U/[(3/2)N k_B]$, we have

$$S = k_B \log \left[\left(\frac{m k_B T}{2 \pi \hbar^2} \right)^{3/2} \frac{e^{5/2}}{N/V} \right]^N \qquad (1.5.2)$$

where we have used $5/2 = \log e^{5/2}$. On comparing this formula with our fundamental equation $S = k_B \log \Gamma$, we see that

$$\Gamma = \left[\left(\frac{m k_B T}{2 \pi \hbar^2} \right)^{3/2} \frac{e^{5/2}}{N/V} \right]^N \qquad (1.5.3)$$

Let us calculate Γ for one mole of helium gas at standard temperature and pressure (STP). The necessary numerical values are listed in Table 1.1.

It is helpful in a calculation like this to start by checking units. The units of the quantity in brackets in Eq. (1.5.3) are basically, $V(m k_B T)^{3/2}/\hbar^3$. We will see later that, when multiplied by 2π, the quantity $\hbar/(m k_B T)^{1/2}$ is an important characteristic length in thermodynamics called the **thermal de Broglie wavelength**. Here we have a volume divided by the cube of this length, so Γ is dimensionless, as it should be.

Table 1.1 *Standard temperature and pressure*

$T = 273.15$ K
$P = 1$ atm $= 760$ mm Hg $= 1 \times 10^6$ dynes/cm^2
Volume of 1 mole of gas at STP $= 22.4$ liters $= 22.4 \times 10^3$ cm^3
Number of atoms in 1 mole $=$ Avogadro's number $= 6.02 \times 10^{23}$
N/V of an ideal gas at STP $=$ Lodschmidt's number $= 2.69 \times 10^{19}$ cm^{-3}
Unit of atomic mass $=$ mass of hydrogen atom or proton $= 1.67 \times 10^{-24}$ g
Atomic mass of helium $= 4 \times 1.67 \times 10^{-24}$ g $= 7 \times 10^{-24}$ g
$k_B = 1.38 \times 10^{-16}$ erg/K
$\hbar = 1.05 \times 10^{-27}$

On substituting numerical values into the quantity in brackets in Eq. (1.5.3) we have

$$\Gamma^{1/N} = \left[\frac{7 \times 10^{-24} \times 1.4 \times 10^{-16} \times 273}{6.3 \times (1.05 \times 10^{-27})^2} \right]^{3/2} \frac{e^{5/2}}{2.7 \times 10^{19^:}} \approx 3 \times 10^6$$

We must still raise this to the Nth power:

$$\Gamma \approx (3 \times 10^6)^{6 \times 10^{23}}$$

It's a bit neater to deal with factors of 10. Since $3 \approx 10^{0.5}$,

$$\Gamma = (10^{6.5})^{6 \times 10^{23}} \approx 10^{39 \times 10^{23}}$$

We said earlier that Γ would be something like 10^N. We see now that it is much bigger than even that. We could not even write down a number like that without repeated exponentiation. Let's consider a different kind of question that depends on these huge numbers.

Imagine a box like the one we have been talking about divided in half by a partition. The partition can't move, but it has a hole in it through which atoms can pass. In thermodynamic equilibrium we expect to find about half the atoms on each side. Let's calculate the probability of finding all the atoms on one side. Since our assumption is that all microscopic states are equally likely, the probability we seek is just the number of states with all the atoms on one side, divided by the total number of states. We can calculate the number of states we want by noticing that in Eq. (1.5.3) $\Gamma \propto V^N$. The number of states with all of the atoms crowded into $V/2$ will have all the other factors in Γ the same. The number we want is therefore

$$\frac{\Gamma(V/2)}{\Gamma(V)} = \left(\frac{1}{2}\right)^N = 2^{-6 \times 10^{23}}$$

Since $2 \approx 10^{0.3}$, the same number can be written as $10^{-1.8 \times 10^{23}}$.

Now, this is a probability. A more useful question might be, how long would we have to watch our box before we see this happen? To find out we must also know how often a gas changes its state. Crudely speaking, that happens every time two atoms collide. A typical value for that might be 10^{-12} seconds. We can say that the gas changes its state 10^{12} times per second. Then the probability per second of finding all of the atoms in one half of the box is

$$10^{12} \times 10^{-1.8 \times 10^{23}} = 10^{12 - 1.8 \times 10^{23}} \approx 10^{-1.8 \times 10^{23}}$$

(The last two numbers aren't really approximately equal, but their exponents differ in the 23rd decimal place.) Obviously, if we want to see this happen, it won't be enough to wait for one second. Suppose we watch the box for a year, roughly $\pi \times 10^7$ seconds $\approx 10^{7.5}$ seconds. Then the probability is

$$10^{12} \times 10^{7.5} \times 10^{-1.8 \times 10^{23}} = 10^{19.5 - 1.8 \times 10^{23}} \approx 10^{-1.8 \times 10^{23}}$$

The age of the Universe is about 10^{10} years. Watching the box that long increases our chances to

$$10^{(29.5 - 1.8 \times 10^{23})} \approx 10^{-1.8 \times 10^{23}}$$

For this purpose it makes no difference whether we express our units in seconds, years or the age of the Universe, we always get the same probability per unit time.

The second law of thermodynamics tells us that the event we have been waiting for will never happen. The example we have given is an exaggerated one, but it tells us why, although the second law is indeed only probably true, it is nevertheless very, very dependable.

Problem 1.8
The logic of the last section also illustrates why it makes no difference whether we disconnect our boxes before computing $\Gamma = \Gamma_1 \Gamma_2 \ldots$. Explain.

Problem 1.9
Two isolated boxes of helium gas, each consisting of one mole at atmospheric pressure, have the same temperature $T_0 = 300$ K. They are briefly put into thermal contact then separated again and allowed to come to equilibrium. Find the probability that one of the boxes will be found to have a temperature $T_1 = 310$ K.

2

The care and feeding of thermodynamic variables

2.1 The thermodynamic variables

In Chapter 1 we made a truly remarkable simplification of nature. For a body of a given U, N and V it's not necessary to know what each of its atoms is doing. If we allow the body to come to equilibrium, all the microscopic information necessary to specify its macroscopic state is contained in the value of a single variable, the entropy. In this chapter we shall explore some of the consequences of that profound insight.

If we know explicitly the equation $U = U(S, V, N)$, then we have enough information to find anything we want. For this reason S, V and N are said to be the ***proper variables*** of the energy U. We have already had some examples of how information can be extracted from this equation, and we shall have some more (Problems 2.5 and 2.7). In reality the equation is hardly ever known explicitly for any real system, but the fact that it exists in principle – which was the point of the first chapter – is what we wish to exploit.

There are seven thermodynamic variables, and these can be grouped into classes in various ways. One way is to assemble them into ***conjugate pairs***. Each pair, when multiplied together, has the units of energy. They can also be classified as either ***extensive*** or ***intensive***. Both are shown in Table 2.1.

The extensive variables are each proportional to the size of the system. The intensive variables do not depend in any way on the size. In fact we saw in the discussions and examples of Chapter 1 that T, P and μ tend to be the same in different parts of a system in equilibrium. The rule is that all three are uniform everywhere in such a system.[1] The U, S, V and N, on

---

[1] In an external force field, like gravity, this statement has to be modified. T is always uniform, but P (and with it μ) may vary to balance external forces.

Table 2.1 *Extensive and intensive variables*

Extensive variables	U, S, V, N
Intensive variables	T, P, μ
Conjugate pairs	$TS, PV, \mu N$

Table 2.2 *Ease of measuring thermodynamic variables*

Easy to measure	T, P, V, N
Hard to measure	U, S, μ

the other hand, are the sums of U, S, V and N of the parts of the body. Each conjugate pair of variables is made up of one extensive and one intensive variable.

Those are the traditional ways of classifying the thermodynamic variables. There's another, more revealing way of dividing them up: into those that are easy to measure and those that are hard to measure. If we are given an arbitrary body in a laboratory, it is relatively easy to find out its temperature, pressure and volume, and even (by weighing it if we know its composition) the number of molecules or atoms in it. It is very much more difficult to determine its energy, its entropy or its chemical potential. This observation is summarized in Table 2.2.

One of the principal purposes of equilibrium thermodynamics is to find ways to substitute easy measurements for hard ones.

2.2 Partial derivatives

To simplify our notation, let us assume until further notice that we are dealing with bodies of fixed N. In that way, N and μ will drop out of our equations. For example, changes of energy between equilibrium states are given by

$$dU = T \, dS - P \, dV \tag{2.2.1}$$

This equation is the differential form of

$$U = U(S, V)$$

Any of our variables can be written as a function of any two others. For example,

$$S = S(P, T)$$

with the consequent differential form

$$dS = \left(\frac{\partial S}{\partial P}\right)_T dP + \left(\frac{\partial S}{\partial T}\right)_P dT \qquad (2.2.2)$$

In an equation like $S = S(P, T)$, P and T are said to be ***independent variables*** (because we are free to choose any value for each of them) and S is the dependent variable (since, having chosen P and T, S is fixed by the equation). Equations like (2.2.1) and (2.2.2) can be used to find expressions for other partial derivatives. Let us see by example how these manipulations are done.

Example 2.2.1
Find an expression for $(\partial U/\partial P)_T$.
    In Eq. (2.2.1) we "divide" by dP, holding T constant (which makes the derivative a partial one):

$$\left(\frac{\partial U}{\partial P}\right)_T = T\left(\frac{\partial S}{\partial P}\right)_T - P\left(\frac{\partial V}{\partial P}\right)_T$$

Remember that T and P themselves are defined as partial derivatives, $T = (\partial U/\partial S)_V$ and $P = -(\partial U/\partial V)_S$.

We can use these manipulations to illustrate a number of properties of partial derivatives that will be useful to us. To start with, suppose we use Eq. (2.2.2) to calculate $(\partial S/\partial V)_T$. We use the same manipulation as in the example above except that, since T is constant, $dT = 0$. We get

$$(\partial S/\partial V)_T = (\partial S/\partial P)_T(\partial P/\partial V)_T$$

The principle we've discovered is that you can break up a partial derivative, i.e. $(\partial S/\partial V)_T$, by "dividing" and "multiplying" by the same differential, dP, so long as you hold constant the other independent variable, T. Since this is a theorem (always true), we shall express it in a more detached way, using variables that don't have specific meanings:

Theorem 1

$$\left(\frac{\partial x}{\partial y}\right)_z = \left(\frac{\partial x}{\partial w}\right)_z \left(\frac{\partial w}{\partial y}\right)_z$$

A real mathematician would have started by saying something like "Let there exist a function $x = x(y, z)$ which is integrable on a domain ...". We will not fuss with such niceties. When we write the symbol $(\partial x/\partial y)_z$ we mean it to imply the existence of the function $x(y, z)$ having the necessary properties. In addition to $x(y, z)$, the expression above implies the existence of $x(w, z)$ and $w(y, z)$.

Two more properties are easily obtained. Consider $P = P(T, V)$, which means

$$dP = \left(\frac{\partial P}{\partial T}\right)_V dT + \left(\frac{\partial P}{\partial V}\right)_T dV \qquad (2.2.3)$$

The same equation could be written $V = V(P, T)$, or, in other symbols,

$$dV = \left(\frac{\partial V}{\partial P}\right)_T dP + \left(\frac{\partial V}{\partial T}\right)_P dT \qquad (2.2.4)$$

Substitute Eq. (2.2.4) into (2.2.3), eliminating dV:

$$dP = (\partial P/\partial T)_V \, dT + (\partial P/\partial V)_T [(\partial V/\partial P)_T \, dP + (\partial V/\partial T)_P \, dT]$$

which can be reorganized into

$$[1 - (\partial P/\partial V)_T(\partial V/\partial P)_T] dP = [(\partial P/\partial T)_V + (\partial P/\partial V)_T(\partial V/\partial T)_P]_T \, dT \qquad (2.2.5)$$

This last equation is identically true; that is, it must hold for any dP and any dT. We can, for example, change P at constant T by changing V. Then the right-hand side is zero because $dT = 0$, but dP is not zero. Therefore the quantity in brackets on the left must equal zero, so we have

$$\left(\frac{\partial P}{\partial V}\right)_T = \frac{1}{(\partial V/\partial P)_T}$$

In other words, a partial derivative can be turned upside down like an ordinary fraction:

Theorem 2

$$\left(\frac{\partial x}{\partial y}\right)_Z = \frac{1}{(\partial y/\partial x)_Z}$$

Going back to Eq. (2.2.5), we could change T while varying V in such a way as to keep P constant. The equation can then be satisfied only if the quantity on the right in brackets is zero:

$$\left(\frac{\partial P}{\partial V}\right)_T \left(\frac{\partial V}{\partial T}\right)_P \left(\frac{\partial T}{\partial P}\right)_V = -1$$

This is known as the chain rule. Each of the three variables in the game appears cyclically in each of the three possible positions in the partial derivatives:

Theorem 3

$$\left(\frac{\partial x}{\partial y}\right)_z \left(\frac{\partial y}{\partial z}\right)_x \left(\frac{\partial z}{\partial x}\right)_y = -1$$

While we are at it, we shall state one more theorem, this time without proof: In a second partial derivative, the order of differentiation is irrelevant. If we have $S = S(P, T)$, then

$$\left[\frac{\partial}{\partial T}\left(\frac{\partial S}{\partial P}\right)_T\right]_P = \left[\frac{\partial}{\partial P}\left(\frac{\partial S}{\partial T}\right)_P\right]_T$$

This notation keeps track of all the variables, but it is cumbersome. As long as we remember that S depends only on T and P, so that if we take a derivative with respect to one we must be keeping the other constant, we may write

$$\frac{\partial^2 S}{\partial T\, \partial P} = \frac{\partial^2 S}{\partial P\, \partial T}$$

Theorem 4

$$\frac{\partial^2 x}{\partial y\, \partial z} = \frac{\partial^2 x}{\partial z\, \partial y}$$

Notice an important special application of Theorem 4. Since

$$\frac{\partial^2 U}{\partial S\, \partial V} = \frac{\partial^2 U}{\partial V\, \partial S}$$

and since $\partial U/\partial S = T$ and $\partial U/\partial V = -P$, we have

$$\left(\frac{\partial T}{\partial V}\right)_S = \left(\frac{\partial P}{\partial S}\right)_V \tag{2.2.6}$$

Equation (2.2.6) is perfectly general, true for any body of fixed N, no matter what it is made of.

Problem 2.1
Using the equation of state of the ideal gas, $PV = Nk_BT$, write the functional forms

$$P = P(T, V)$$

$$V = V(P, T)$$

Using these equations, evaluate all of the partial derivatives in Eq. (2.2.5) and show that the quantities in brackets on each side are equal to zero.

Problem 2.2
Show that

$$\left(\frac{\partial T}{\partial V}\right)_S = -\frac{(\partial V/\partial T)_P (\partial P/\partial V)_T}{(\partial S/\partial T)_V}$$

Problem 2.3
Show that

$$\left(\frac{\partial U}{\partial T}\right)_V = -T\left(\frac{\partial V}{\partial T}\right)_S \left(\frac{\partial P}{\partial T}\right)_V$$

2.3 Some special partial derivatives

Even with only the five variables at present on the board, there are $5 \times 4 \times 3 = 60$ partial derivatives one can write. If we include more variables there are many more partial derivatives. These can be connected by an almost endless variety of relations. It is necessary to keep a clear head in plowing through this mathematical blizzard. Still, it is worth the trouble because the rewards are considerable. We cannot use manipulations of partial derivatives to find the "hard-to-measure" quantities listed in Table 2.2, but what we can often do is to deduce changes in hard-to-measure quantities from changes in quantities that are easy to measure.

Of all the partial derivatives possible, a few are singled out for special recognition and given names of their own. We've already seen that in the cases of T, P and μ. Three other such quantities are

$$C_V = T(\partial S/\partial T)_V \tag{2.3.1}$$

which is called the **heat capacity at constant volume**,

$$K_T = -\frac{1}{V}\left(\frac{\partial V}{\partial P}\right)_T \tag{2.3.2}$$

which is called the **isothermal compressibility**, and

$$\beta = \frac{1}{V}\left(\frac{\partial V}{\partial T}\right)_P$$

which is called the **thermal expansion coefficient**.

All three of these should be regarded as measurable quantities. For most substances the relation involving P, T and V for a given N is among the most available of thermodynamic data. An equation of the form $P = P(T, V)$ is called the **equation of state**. In sorting out our variables, we should regard the equation of state and its derivatives, including K_T and β, as known or at least measurable quantities.

Next consider C_V. Suppose the substance we are interested in fills a rigid can (so that its volume will be constant), and it is isolated from the rest of the world, except for a coil of electric heater wire, which is wound onto the can. If we pass a current through the wire we are putting energy into the system. The energy will be $IV_0 t$, where I is the electric current, V_0 is the voltage across the wire and t is the time for which the current runs. We certainly expect that operation to make the temperature of the stuff in the can go up. If we measure the change in temperature, ΔT, then we have a way to estimate one of our partial derivatives:

$$C_V = T(\partial S/\partial T)_V = (\partial U/\partial T)_V = IV_0 t/\Delta T$$

The equality becomes exact in the limit of small ΔT. Thus we have found a way to measure C_V (that is exactly how C_V is actually measured). That's important because for the first time it gives us a way to at least measure changes in the elusive quantity S.

Example 2.3.1

Find out how $(\partial T/\partial V)_S$ may be obtained from measurable quantities.

Solution.

From Problem 2.2 we have

$$\left(\frac{\partial T}{\partial V}\right)_S = -\frac{(\partial V/\partial T)_P(\partial P/\partial V)_T}{(\partial S/\partial T)_V}$$

$$= \frac{-\beta V(1/(VK_T))}{C_V/T}$$

$$= \beta T/(C_V K_T)$$

Most substances expand when they are heated. That is to say, β is usually positive. However, there is no thermodynamic reason why it must be positive and there exist substances (for example, liquid helium at certain temperatures and pressures) that contract when they are heated at constant pressure. Thus β can sometimes be negative. On the other hand, C_V and K_T are never negative for any substance in equilibrium. We leave the proof of that statement to Problems 2.11 and 2.12. However, formal proof aside, this assertion expresses facts that are easy to understand intuitively.

When we described how to measure C_V we said that the temperature would go up if we added energy to our sample at constant volume. That means $C_V = (\partial U/\partial T)_V$ is positive.

For K_T to be positive $(\partial V/\partial P)_T$ must be negative. Let us imagine what would happen if for some substance $(\partial V/\partial P)_T$ were positive instead. That would mean any increase in V makes P increase, which makes V increase more and so on. Such a substance would be explosively unstable, and so could not exist for long in nature.

Two more quantities that have names of their own are

$$C_P = T\left(\frac{\partial S}{\partial T}\right)_P \qquad (2.3.3)$$

which is called the **heat capacity at constant pressure**, and

$$K_S = -\frac{1}{V}\left(\frac{\partial V}{\partial P}\right)_S$$

which is called the **adiabatic compressibility**.

C_P and K_S are not themselves easily measured, but they may be deduced from other quantities that are. For C_P start with $S = S(T, V)$. Then,

$$dS = \left(\frac{\partial S}{\partial T}\right)_V dT + \left(\frac{\partial S}{\partial V}\right)_T dV = \frac{C_V}{T} dT + \left(\frac{\partial S}{\partial V}\right)_T dV$$

and

$$C_P = T\left(\frac{\partial S}{\partial T}\right)_P = C_V + T\left(\frac{\partial S}{\partial V}\right)_T \left(\frac{\partial V}{\partial T}\right)_P = C_V + TV\beta\left(\frac{\partial S}{\partial V}\right)_T$$

To get rid of $(\partial S/\partial V)_T$ we use the chain rule,

$$\left(\frac{\partial S}{\partial V}\right)_T = -\left(\frac{\partial S}{\partial T}\right)_V \left(\frac{\partial T}{\partial V}\right)_S = -\frac{C_V}{T}\left(\frac{\partial T}{\partial V}\right)_S$$

We have already evaluated $(\partial T/\partial V)_S$ in Example 2.3.1: $(\partial T/\partial V)_S = -\beta T/(C_V K_T)$, so $(\partial S/\partial V)_T = \beta/K_T$, and, finally,

$$C_P = C_V + \frac{TV\beta^2}{K_T} \tag{2.3.4}$$

By rather similar manipulations, you can show that

$$K_S = \frac{C_V}{C_P} K_T \tag{2.3.5}$$

Problem 2.4
Estimate numerically $(\partial S/\partial V)_U$ for a block of copper sitting on your desk.

Problem 2.5
An atomic ideal gas obeys Eq. (1.1.4),

$$U = \frac{3}{2} N k_B \left(\frac{N}{V}\right)^{2/3} \exp\left[\frac{S}{(3/2)Nk_B} - s_0\right]$$

Find C_V, C_P, K_T and β.

Problem 2.6
For the ideal gas in Problem 2.5, show that Eqs. (2.3.4) and (2.3.5) are satisfied.

Problem 2.7
A certain substance has $C_V = aVT^3$. At $T = 0$, the entropy is zero, and U is independent of V and equal to zero.

(a) Find $U(T, V)$.
(b) Find PV/U (a pure number).
(c) Find the equation of state.

Problem 2.8
Show that C_V and C_P can also be written

$$C_P = C_V + TVK_T[(\partial P/\partial T)_V]^2$$

Problem 2.9
Derive the (rather useless) formula

$$\left(\frac{\partial P}{\partial T}\right)_S = \frac{\beta + \frac{1}{V}\left(\frac{\partial S}{\partial P}\right)_V}{K_T}$$

Problem 2.10
Derive Eq. (2.3.4),

$$K_S = \frac{C_V}{C_P} K_T$$

Problem 2.11

Prove $C_V \geq 0$ for any system in equilibrium. Hint: For the system to be in equilibrium, the entropy must be at a maximum. That implies conditions on the first and second derivatives of the entropy. Use that to construct this proof.

Problem 2.12

Prove $K_T \geq 0$ for any system in equilibrium. The proof is analogous to Problem 2.11.

Problem 2.13

Prove C_P and K_S are always ≥ 0.

Problem 2.14

A certain substance has a negative thermal expansion coefficient. If its pressure is increased at constant volume, will its temperature go up or down?

Some solutions are as follows.

2.4 $dU = T\, dS - P\, dV$, so

$$(\partial S/\partial V)_U = P/T = 1 \text{ atm}/300 \text{ K} = 3.3 \times 10^3 \text{ dynes/cm}^2 \text{ K}$$

2.9 $V = V(T, P)$; $dV = (dV/dT)_P\, dT + (dV/dP)_T\, dP$, so $(1/V)(\partial V/\partial T)_S = \beta - K_T\, (\partial P/\partial T)_S$, but $(\partial V/\partial T)_S = -(\partial S/\partial P)_T$, so $-(1/V)(\partial S/\partial P)_T = \beta - K_T\, (\partial P/\partial T)_S$ or, finally,

$$(\partial P/\partial T)_S = \frac{\beta + (1/V)(\partial S/\partial P)_T}{K_T}$$

2.14 Chain rule:

$$\beta = (1/V)(\partial V/\partial T)_P = (1/V)[-(\partial P/\partial T)_V(\partial V/\partial P)_T] = K_T(\partial P/\partial T)_V$$

Since $K_T \geq 0$ if $\beta < 0$, $(\partial P/\partial T)_V$ must be negative. T goes down.

2.4 Thermodynamic operations

Each of the partial derivatives we have been studying should be read as a symbolic description of a physical process. Thus, for example, when we write $(\partial P/\partial T)_V$ we are asking: If we warm a body at constant volume, how much will its pressure increase? In some cases the operation described can be performed in a laboratory, or occurs in nature. In other cases, the operation is impractical, but can be imagined in a thought experiment. In all cases we should be aware of precisely what is being described.

In every instance we are supposing the existence of a functional relationship among three or more variables, say, $P = P(T, V)$. We then imagine holding one of the independent variables constant (say, V), changing the other (T) by a known amount, and measuring the consequent change in the dependent variable (P). The functional relationship does not exist in the first place unless the body is in equilibrium, so we must always allow our body to reach thermodynamic equilibrium before and after our operation. With that in mind the question becomes, how do we go about changing one quantity while keeping another constant?

It's not hard to imagine keeping the volume of a sample constant, particularly if it's a liquid or a gas. We simply put it in a rigid container. Processes that occur at constant volume are said to be *isochoric*.

If we warm a solid body in the air, the pressure is automatically constant and equal to atmospheric pressure. On the other hand, warming a fixed quantity of gas at constant pressure requires some planning. We can do it in a container in the form of a cylinder with a movable piston. The atmosphere will do the job, keeping the pressure fixed at one atmosphere plus the weight of the piston divided by its area applies the external force on the piston. In principle, we could use this arrangement to measure C_P, the heat capacity at constant pressure, but that is unnecessarily complicated and it is seldom done that way. Any process that takes place at constant pressure is said to be *isobaric*.

To keep the temperature constant, we place our body in contact with a much larger body whose temperature is constant. The larger body is called a heat bath or a temperature bath. Its most important property is that it's so big that its temperature is not measurably affected by changes in the temperature of our puny sample. Processes at constant temperature are said to be *isothermal*.

In reality there is no such thing as perfect isolation. Energy is transmitted, even through a vacuum, by electromagnetic radiation. An isolated system is always an idealization, like the massless strings and frictionless pulleys of mechanics. For that matter there is no such thing as perfectly constant volume or temperature or pressure. We use such idealizations because they help us sort things out, because there is no intrinsic limit to how closely we can approach them, and because they are often an excellent approximation to what really happens.

If the energy of a body does change, the way we formulated the problem in Chapter 1 led us naturally to divide changes into two parts, as expressed in Eq. (2.2.1),

$$dU = T\, dS - P\, dV$$

Any energy transferred without changing the volume of the body is said to have flowed in the form of **heat**, Q:

$$dQ = T\, dS$$

If the body decreases its energy purely by increasing its volume, it is said to have done **work**, W:

$$dW = P\, dV$$

The idea that a single quantity, energy, can be transferred either as heat or work is one of the cornerstones of the discipline. It's called **the first law of thermodynamics**. We will look more closely at the first law later. First let us think about the operations necessary to transfer pure heat or to do pure work.

We have described an experiment in which a can of stuff was heated electrically at constant volume. That is a perfect example of a pure transfer of heat. We can also imagine an example in which a body expands when it is heated. In that case $dV \neq 0$ and so work is also done. If the volume does not change, the energy transferred is equal to the heat transferred. Then

$$C_V = T(\partial S/\partial T)_V = (\partial Q/\partial T)_V = (\partial U/\partial T)_V$$

If, instead, we transfer heat at constant pressure (i.e. isobarically), we have

$$C_P = T(\partial S/\partial T)_P = (\partial Q/\partial T)_P$$

but it is not equal to the derivative at constant P of the energy. Instead we have

$$(\partial U/\partial T)_P = C_P - P(\partial V/\partial T)_P$$

This equation, which follows from Eq. (2.2.1), means that some of the heat put in at constant pressure is converted to work done on the source of the constant pressure.

The next question is can we perform work without transferring heat? The answer is yes, if we ignore details like friction and viscosity. If we allow a body to expand without any transfer of heat, work is done, and the process is said to have been **adiabatic**. If in addition no work is turned into heat by friction or viscosity, the entropy doesn't change and the process is said to have been **isentropic**.

Notice that, whenever we speak of heat or work, we are always speaking of a process or a transfer, or in other words of a change. Once a body has come to equilibrium, it contains neither heat nor work. It has only energy.

Let us now see whether we can understand the division between heat and work from a microscopic point of view. To think about it, we'll use the one example we understand in detail, namely the perfect gas. In Section 1.2 we saw that the perfect gas problem could be described as follows: We have N atoms in a box of volume $V = L^3$ with total energy

$$U = B \frac{h^2}{2mV^{2/3}}$$

where $h^2/(2mV^{2/3})$ is the quantum unit of energy for an atom in the box, and B is the number of units of energy to be divided up among the N particles. In the discussion and problems of that section, especially Problem 1.4, we also saw that the number of states of the system, Γ, depended only on N and B, not on m or V, which govern only the size of the unit of energy. If we add energy to the box without changing V, we must clearly be changing B. That just as clearly changes Γ and hence S. This is what we mean by heat.

Is it possible to change U without changing S? Suppose for a moment that the gas is in one particular quantum state, using up all N particles and B units of energy. Now let us increase the volume. The unit of energy gets smaller. Levels of different energy draw closer together. But each level retains its identity. If, before the operation, a certain set of occupation numbers used up N atoms and B units of energy, after the operation the same set of occupation numbers will still use up N atoms and B units of energy. Changing V all by itself does change the total energy U because the unit of energy changes, but B and N remain as they were. The entropy S, as we have just argued, depends only on N and B, so it does not change. We have succeeded in changing U at constant S. That change is work.

Problem 2.15
Using the logic of the argument above, prove that the pressure of a perfect gas is always given by $P = (2/3)U/V$. We found the same result in Example 1.1.2, using the equations of the ideal gas. But the ideal gas is a special case of the perfect gas, valid at high T and low density. Here you are demonstrating a more general result.

The ideal gas obeys the equation

$$PV = Nk_BT \qquad (2.4.1)$$

On combining this with the result of Problem 2.15 we have

$$U = \frac{3}{2}Nk_BT \qquad (2.4.2)$$

Let's apply our new knowledge to analyze three different ways to expand an ideal gas.

1. An isolated system consists of a box with a partition dividing it into two parts. The gas starts in one part with the other part evacuated. The partition is then broken, allowing the gas to expand into the other part of the box.

   In this process the energy of the system is constant. No work is done because the gas expands against zero pressure, and the system is isolated so no energy gets in or out. This process is called a *free expansion*.
2. The gas, in a cylinder, pushes out a piston doing work. The whole arrangement is sunk in a temperature bath, so it is done at constant T. This process is called an *isothermal expansion*.
3. Cylinder and piston again, so work is done, but this time the system is isolated. This process is called an *adiabatic expansion*.

In a free expansion of an ideal gas, since $U = (3/2)Nk_BT$ and U is constant, so is T. The entropy must increase when V does, because $dU = T\, dS - P\, dV = 0$, so

$$\Delta S = \int \frac{P}{T}\, dV \text{ at constant } T$$

$$= \int \frac{Nk_B}{V}\, dV$$

(we have used $PV = Nk_BT$) and

$$S_{final} - S_{initial} = Nk_B \log(V_{final}/V_{initial})$$

Notice that, although the gas is certainly not in equilibrium during a free expansion, we can still use our equilibrium thermodynamics to find out how much quantities like T and S change. That is because we have kept track of the total energy of the system as well as its volume, and we know that the state of the system is a unique result of those two quantities when it is in equilibrium before the event occurs and after it is over.

In an isothermal expansion, T is constant and thus, for an ideal gas, U is again constant. Hence the entropy changes by the same amount as above. An isothermal expansion differs from a free expansion, however, in that work is done. The work comes not from the internal energy of the gas, which remains constant, but from the bath that is keeping T constant. *In an isothermal expansion of an ideal gas, heat is extracted from the bath and turned into work.* The work done is given by

$$W = \int P \, dV = \int \frac{NkT}{V} \, dV = Nk_{\mathrm{B}}T \, \log(V_{\mathrm{final}}/V_{\mathrm{initial}})$$

Work is also done in an adiabatic or isentropic expansion. In this case, however, there is no external source of heat so the energy of the gas must decrease and so must its temperature. The following argument will clarify what is happening. Write the entropy of the gas as

$$S = S(P, V)$$

with

$$dS = \left(\frac{\partial S}{\partial P}\right)_V dP + \left(\frac{\partial S}{\partial V}\right)_P dV$$

$$= \left(\frac{\partial S}{\partial T}\right)_V \left(\frac{\partial T}{\partial P}\right)_V dP + \left(\frac{\partial S}{\partial T}\right)_P \left(\frac{\partial T}{\partial V}\right)_P dV$$

where we have twice used Theorem 1 of Section 2.2. We substitute in $C_V = (1/T)(\partial S/\partial T)_V$ and $C_P = (1/T)(\partial S/\partial T)_P$ and evaluate the other derivatives from $PV = Nk_{\mathrm{B}}T$. We find

$$dS = C_V \frac{dP}{P} + C_P \frac{dV}{V}$$

But we are interested in a process for which $dS = 0$. Defining $\gamma = C_P/C_V$, we have

$$dP/P = -\gamma \, dV/V$$

or

$$d\log P = -d\log V^\gamma$$

and, finally,

$$PV^\gamma = \text{constant}$$

In an isothermal expansion we have $PV = \text{constant}$ with a different constant (equal to $Nk_{\mathrm{B}}T$) for each temperature at which an expansion takes place. In an adiabatic expansion we have $PV^\gamma =$ constant with a different constant for each entropy at which an expansion takes place. In the adiabatic case the temperature is decreasing as the gas expands, causing P to fall faster than in the isothermal case as V increases. The difference is expressed by the exponent γ. For an atomic ideal gas $\gamma = 5/3$. For a gas of diatomic molecules (such as air) it is typically $7/5$. It is usually constant for a given gas.

Problem 2.16
We have seen that an ideal gas can undergo a free expansion, an isothermal expansion or an adiabatic expansion. We can also imagine it expanding isobarically. For each of those four cases, compute the derivatives $\partial U/\partial V$, $\partial T/\partial V$ and $\partial S/\partial V$.

Problem 2.17
An ideal gas goes through the four expansions in Problem 2.16 starting from the same state. Sketch a plot of pressure versus volume and show the trajectory followed in each kind of expansion (we are interested in the relative slopes of the curves rather than precise quantitative details).

Problem 2.18
If one mole of an ideal gas starts out at STP and goes through each of the four expansions of Problem 2.16 until its volume is doubled, find in each case the work done by the gas, the change in entropy of the gas and the final T and P of the gas.

Problem 2.19
Two chambers, each with a piston, are separated by a fine nozzle, as shown in Fig. 2.1. The nozzle permits gas to leak through slowly. The combined system is isolated except for the forces applied to the piston rods. The experiment starts with the piston on the right up against the barrier, and a quantity of gas in the left-hand chamber at pressure P_1 and

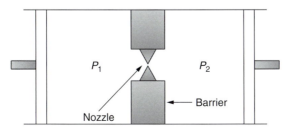

Figure 2.1 Two chambers separated by a fine nozzle.

volume V_1. Now both pistons are moved to the right in such a way that the pressure on the left is maintained at a constant value P_1 and the gas on the right is maintained at (lower) pressure P_2. The gas squirts through the nozzle (squirting is not an equilibrium process) until all the gas is in the right-hand chamber, occupying volume V_2. If the gas initially had energy U_1 and at the end has energy U_2 show that

$$U_1 + P_1V_1 = U_2 + P_2V_2$$

We shall later call the quantity $U + PV$ the **enthalpy**. The process we have described, which is called a ***Joule–Thomson expansion***, is one in which the enthalpy is conserved.

Problem 2.20
Find the change in temperature of an ideal gas in a Joule–Thomson expansion.

3

Gases and other matters

3.1 The states of matter

The following is a thought experiment. Nothing written here should be construed as instructions for practical action. This is not a homework problem.

You go to your freezer, take out an ice cube, and put it in an experimental fluid of your choice. Some of the choicest fluids for this experiment are made along the banks of glens in Scotland. A scientist often feels compelled to repeat an experiment many times to be sure that the results are dependable. If you do so in this case, you might not be able to remember the results.

Here is what you observe. The fluid starts out at room temperature. The ice is initially at the temperature of the freezer, say, 260 K or −13 °C. At first the ice warms up, cooling the fluid. But then, when its temperature reaches 273 K or 0 °C something happens that is both extraordinary and dramatic. The ice refuses to get any warmer. Whereas just before it had no difficulty raising its temperature as it absorbed heat, it now absolutely refuses to warm the slightest bit more. Instead it undergoes a catastrophic change in its microscopic state. As it absorbs more heat from its surroundings, the molecules separate from their solid crystal structure. The ice melts into water. The whole profound transformation takes place without permitting any change in temperature until it is completely finished. If that were not the case, you wouldn't cool your drinks with ice. You might as well throw cold rocks in them.

The melting of ice, like the boiling of water, is a ***phase transition***. Phase transitions are among the most remarkable and interesting phenomena in nature. You might expect that a phase transition occurs whenever any kind of matter changes from one to another of its principal states – solid,

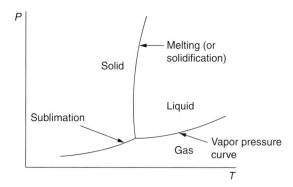

Figure 3.1 A *P–T* diagram.

liquid or gas. In fact the story is not so simple. We have already seen that, for any given quantity of matter, its thermodynamic equilibrium state is uniquely determined by specifying any two of its thermodynamic variables. Thus, for example, at $T = 80$ K and $P = 1$ atm, one mole of argon has a definite volume, energy, entropy and so on. It is also in a unique phase (in this case argon is a solid). If we imagine a plane, with pressure plotted along one axis and temperature along the other, it should be possible to divide the entire plane into regions where argon is either solid, liquid or gas. Such a plot, called a *P–T* diagram, is shown in Fig. 3.1.

In the $P-T$ plane, the regions in which the various phases exist share common borders called *coexistence curves*. For example, solid and liquid coexist only along a curve called the *melting curve* (or *solidification curve*, depending on how you look at it). The fact that it is a curve (rather than a region) means that, if solid and liquid are to coexist at a given pressure, they can do so only at a unique temperature where the curve crosses that pressure. The ice we started out with warmed up under a pressure of one atmosphere (applied by the atmosphere) until it reached its melting curve at 273 K, whereupon it had to change into water before it could get any warmer. The curves separating a solid or a liquid from a gas always have a positive, concave upward slope. The melting curve can slope forward as in Fig. 3.1 or occasionally backward, as it does for water. We will delve further into the cause and significance of these details later.

The curves that separate liquid, solid and gas intersect at a unique point called the *triple point*. Only at that pressure and temperature can three phases coexist simultaneously in thermodynamic equilibrium. The Kelvin temperature scale is determined by defining the temperature of the triple point of water to be

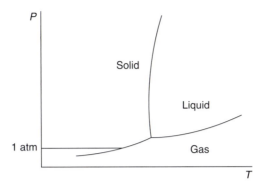

Figure 3.2 Here 1 atm crosses the solidification curve, as it does for carbon dioxide.

$$T_3 = 273.16 \text{ K} \tag{3.1.1}$$

It is convenient to use the triple point of water as the standard of temperature because, in principle, it can be reproduced in any laboratory anywhere in the world.

The point at which the vapor pressure curve (of water or any other substance) intersects one atmosphere is called the ***normal boiling point*** (NBP). One cannot find the normal boiling point on a diagram like Fig. 3.1 that does not have a specific scale because the NBP is not an intrinsic property of matter. It is the constant temperature at which the liquid boils if it is kept at an arbitrarily chosen pressure. The particular pressure we've chosen is valid only on our own planet and even here the pressure is often a little more or a little less (at high altitudes considerably less) than one atmosphere (atm). Moreover, a substance need not even have an NBP if its vapor pressure curve does not cross 1 atm. At that pressure, carbon dioxide *sublimes*, that is it evaporates directly from the solid as sketched in Fig. 3.2.

The peculiar numerical value of Eq. (3.1.1) is chosen in order that there will be exactly 100 kelvins between the triple point and the NBP of pure water. In other words, the temperature interval 1 K has been chosen to be equal to 1 °C (one degree Celsius). The numerical value of Boltzmann's constant, $k_B = 1.38 \times 10^{-16}$ erg/K, is also a consequence of that choice.

In addition to the triple point, there is one other point on the *P–T* diagrams we are studying, which is called the ***critical point***. The critical point is the end point of the vapor pressure curve. It occurs at a unique temperature and pressure for each substance.

To understand what happens when the end point of the liquid–gas coexistence is reached, imagine the two phases in equilibrium at some point along the curve. The liquid, of course, is the denser of the two. If the stuff is in a sealed transparent capsule, you can clearly see the interface between the liquid on the bottom of the capsule and the gas on top. If we raise the temperature, the system follows the vapor pressure curve, and the liquid and vapor stay in equilibrium but at higher pressure. The increased pressure increases the density of the gas, while the increased temperature tends to decrease the density of the liquid. Thus the difference in density between the two phases becomes smaller. This trend continues as we move along the curve until there is no difference at all. Beyond that point there is no interface; nor is there any distinction between the phases.

The observation we have just described was first made in ether, by Charles Cagniard de la Tour (1777–1859). Its real significance was first grasped by the great intuitive genius Michael Faraday (1791–1867), who called it the disliquefying point. Faraday understood that, beyond that point, a given substance would never turn into a liquid no matter how much it was compressed. The term critical point was contributed in a lecture at Great Britain's Royal Institution by Thomas Andrews (1813–1885). The existence of the critical point raises some tricky questions. Is the stuff beyond the critical point a liquid or a gas? In fact, what is the difference between a liquid and a gas?

Liquids are denser than gases, to be sure, but that doesn't necessarily help us. Suppose we have a capsule filled with a homogeneous fluid. Will knowing

Michael Faraday

Thomas Andrews

its density help us decide whether it's a gas or a liquid? At what numerical density do gases cease to exist and liquids begin? On the other hand, it would be silly to pretend that there is no difference or that we don't know what the difference is. Any child can tell a liquid from a gas, and we sophisticated students of thermodynamics ought to be able to distinguish between them too. Imagine you have just jumped off a diving board into a swimming pool. You are poised in mid-air just above the surface of the water. Do you know whether you are in a gas or in a liquid? You will know when you leave the gas and enter the liquid, and with a splash you do.

Now, while you are gurgling under the water, we will, for purposes of thermodynamic clarity, replace the air above the pool with pure water vapor. In the meantime, there is no doubt at all that you are in a liquid. At this point we go through the following procedure: We heat you up to get around the critical point of water, then cool you down again and bring you back to the original pressure. The path we followed is shown in Fig. 3.3.

You are now once again poised in mid-vapor, just above the surface, presumably little the worse for having gone above the critical temperature of water, 697 K, and the critical pressure, 221 atm. You are now unmistakably back in the gas. There was no doubt about when you splashed into the liquid from the gas, but now you have passed back into the gas in a different way. At what point in your ordeal did the change take place? At what point along the path could you have said: "Aha. Here is where

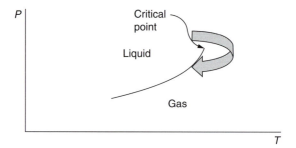

Figure 3.3 A path around the critical point.

I cease to be in the liquid and return to the gas?" There is no such point.
You have moved from the liquid to the gas without ever passing through
a phase transition. That means there must be no fundamental difference
between a gas and a liquid. They differ only in density, a quantitative
difference that can take on any value, including zero.

If you examine how you intuitively distinguish a liquid from a gas, you
will find that it is always associated with the presence of a free surface, that
is, an interface between liquid and gas. When we say that a liquid sloshes,
we are describing the motion at the interface. When you splashed into the
swimming pool, you were passing through the surface of the water. We
can tell the difference between a liquid and a gas only when the two exist
together. It is precisely when a liquid and vapor of the same substance
coexist that there is indeed a phase transition between them, and in fact the
interface is the physical place where the phase transition occurs.

Although we cannot always be sure whether a substance is a liquid or a
gas, we can always uniquely distinguish an equilibrium solid. It seems
obvious enough that solids are rigid; they do not flow. But that property
is not sufficient for the distinction we wish to make. There are liquids
that flow easily, like water, others that flow more slowly, like honey, and
materials that flow very slowly indeed, like glass. Rigidity, like density, is
a quantitative property, one that can change continuously. If rigidity
were the only difference between a liquid and a solid, we could imagine
changing smoothly from one to the other, as we did in going from liquid
to gas. The difference between water and ice, however, is of a more
fundamental nature. When water freezes, the molecules, instead of behaving chaotically like a dense gas, organize themselves into a crystal lattice.
The lattice is a geometrical structure that repeats itself throughout the
space occupied by the lattice. Every substance, when it freezes in equilibrium, forms a crystal with a definite lattice.

A crystal lattice has special geometric properties called ***symmetries***. The simplest and most universal symmetry is this one: If you start at some particular atom and march through the crystal in a straight line in any direction, you would encounter other atoms of the same kind at definite intervals. If you started at the same atom and marched in the opposite direction you would encounter atoms of that kind at exactly the same set of intervals. Using this kind of symmetry as a criterion, a microscopic being inside a material could always tell whether or not it was inside a crystalline solid. As a practical matter, instead of sending microscopic beings into a material, we can use a beam of X-rays. The X-rays are diffracted by the atoms of the material, emerging in directions that reflect the internal symmetry, if any exists. Thus it is possible in principle to uniquely determine whether or not a material is a solid.

Let us compare this discussion with what we said earlier about the difference between a liquid and a gas. In that case, because there's a critical point at which the coexistence curve terminates, we could find ourselves going from one phase to the other without being able to identify the instant at which the change took place. Now we have the opposite situation. Since we can always distinguish between a liquid and a solid, the kind of trip we took going from liquid to gas cannot be possible in going from liquid to solid. That appears to mean that there cannot be a critical point where the solid–liquid coexistence ends. Is that correct? There are so many varieties of matter, nature is so rich in possibilities, that it's hard to find simple general statements about the behavior of matter that are always true. Nevertheless, this one is correct. Melting curves never end in critical points. There is no way to pass from a solid to a liquid without going through a phase transition.

Both our reasoning and our conclusions apply, of course, only to crystalline solids. There are rigid materials that are not crystalline. Glass is an example, and there are others called amorphous solids. When glass is heated it softens rather abruptly, but it does not melt in the sense of going through a phase transition. Materials like glass are interesting and important in their own right, but they are not equilibrium states of matter. Glass cannot be found in the P–T phase diagram of SiO_2 of which it is composed. If glass is heated and then cooled slowly, so that it may come to equilibrium, it crystallizes into quartz. The general rule we have laid down applies only to crystalline solids, but it applies to all of them.

Up to now, we have focused our attention on the pressures and temperatures at which the various states of matter can be found. There are other revealing ways to look at the same information. A particularly

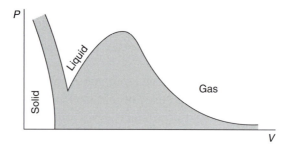

Figure 3.4 A phase diagram in the *P–V* plane.

important combination is the pressure and volume of a given quantity of matter. At a given pressure and temperature, matter is always in one state or another except on the coexistence curves. The same statement cannot be made, however, for a given pressure and volume.

Consider a liquid just at its vapor pressure. Since the liquid–gas coexistence curve in the *P–T* plane has no width, the liquid can be changed entirely into gas without any change in pressure or temperature. Since one mole of gas occupies more volume than one mole of liquid at the same *P* and *T*, it is clear that in a plot of *P* versus *V* there must be a gap between the volume of the liquid and the volume of the gas at the same pressure. Volumes in between those extreme values are occupied by one mole of the same stuff when it is partly liquid and partly gas. Thus in the *P–V* plane there are not coexistence curves, but rather coexistence regions.

A *P–V* phase diagram for a typical substance is shown in Figure 3.4. At small *V* (i.e. high density), the material is always solid. At sufficiently large *V* it is always gas. In between there is a region of liquid. The shaded regions are those where two or more phases coexist. The best way to understand the significance of this graph is to follow a substance through its various phases as we compress it from large volume to small at constant temperature. A constant-temperature path is called an ***isotherm***.

First let us see what is going on by following the action in the already familiar *P–T* plane, in Fig. 3.5.

Choose a temperature between the triple point and the critical point. An isotherm in this plane is simply a vertical line. Starting at large *V*, which means low *P* in the gas, *P* rises as we decrease *V*, until we encounter the vapor pressure curve. Being an ideal or almost ideal gas up to that point, it will obey, or nearly obey, the ideal gas law, PV = constant (at constant *T*). What we have up to now looks like Fig. 3.6.

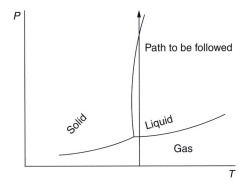

Figure 3.5 An isotherm in the *P–T* plane.

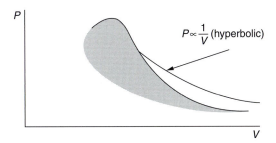

Figure 3.6 A path in the *P–V* plane.

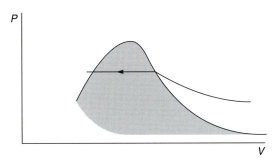

Figure 3.7 Liquefaction.

Now, as we continue to decrease V (you can imagine the stuff to be in a cylinder with a piston we can push in, the whole thing being in a bath at constant T), the gas condenses into a liquid. Condensation takes place at constant pressure, so the path looks like Fig. 3.7.

Once the stuff is entirely liquid, it becomes much more difficult to compress. That means a large increase in pressure is necessary to produce a small change in volume. The path turns sharply upward, as shown in Fig. 3.8.

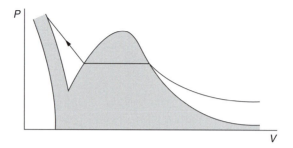

Figure 3.8 The path continues in the *P–V* plane.

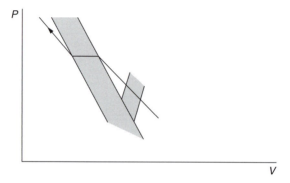

Figure 3.9 Another bit of the path.

At the solidification curve we once again change at constant *P* to a state of higher density (lower *V*), and, when the material is completely solid, it once again becomes hard to compress (Fig. 3.9).

In Fig. 3.10 we show the complete isotherm we have been following, along with four others. Starting below the triple point, curve 1 jumps directly from gas to solid. Curve 2 touches all three phases: its temperature is the triple point. Curve 3 is the one that we have followed. Curve 4 has no horizontal section, but rather an inflection point (of zero slope) that just touches the top of the coexistence region. Its temperature is T_C, the critical temperature. Finally, curve 5 is supercritical. Although it wiggles a little, it suffers no phase transition until it finds the solidification curve at very high pressure.

There are a number of variations on this theme (see Problem 3.4), but these are the basic ideas. Quite aside from topological features – the scales, that is – the characteristic temperatures and pressures vary from one substance to another. A few examples are given in Table 3.1.

Table 3.1 *Characteristic temperatures and pressures*

	Triple point		Critical point		
Substance	T (K)	P (atm)	T (K)	P (atm)	V (cubic cm/g)
Helium, ^{4}He	No solid–liquid– gas triple point		5.19	2.25	14.5
Hydrogen, H$_2$	13.84	0.0695	32.98	12.76	31.8
Neon, Ne	24.6	0.426			
Nitrogen, N$_2$	63.2	0.124	126.2	33.5	

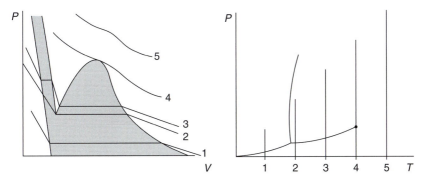

Figure 3.10 Two versions of the phase diagram.

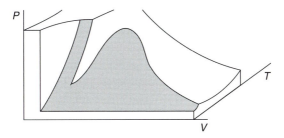

Figure 3.11 A typical *P–T–V* surface.

The graphs that we have been discussing are geometric representations of the fundamental fact of equilibrium thermodynamics: that, for a given amount of a given substance in equilibrium, any of the thermodynamic variables is a unique function of any two others. For example, the geometric expression of the equation of state, $P = P(T, V)$, is that in a coordinate system whose axes are *P*, *T* and *V* the equilibrium states of the substance fall on some definite surface. The surface is sketched in Fig. 3.11.

Josiah Willard Gibbs

The idea of representing the states of matter this way was first pro-posed by the great thermodynamicist Josiah Willard Gibbs. The first American to be awarded a doctorate in mechanical engineering (from Yale University), Gibbs was also one of the first important scientists produced in the New World (two of the others were Benjamin Franklin and Joseph Henry). He was an affable if reclusive man, content to spend his life in New Haven, Connecticut, working for Yale, where his father had been a professor before him. Yale even paid him a small salary, but only after Johns Hopkins had offered him a much larger one in a futile attempt to lure him away.

Although he preferred to work in isolation far from the centers of scientific activity, Gibbs did not neglect to send copies of his results to his colleagues in Europe. There he found supporters, among them a true giant in the history of physics, James Clerk Maxwell (1831–1879). Max-well was a man of irrepressible enthusiasm. Reacting with characteristic zeal to the ideas that arrived from America, he made plaster casts of Gibbs surfaces representing the states of matter and gave them to friends, sending one to Gibbs himself. That one eventually wound up in America's national attic, the Smithsonian Institution in Washington D.C. Photo-graphs of it are shown in Figs. 3.12 and 3.13.

James Clerk Maxwell

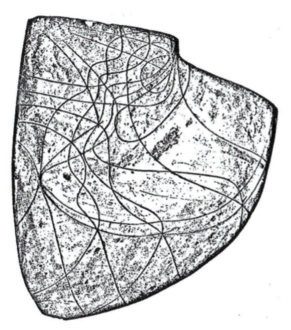

Figure 3.12 A photo of a cast of the Gibbs surface. Its size is roughly 15 cm on each edge.

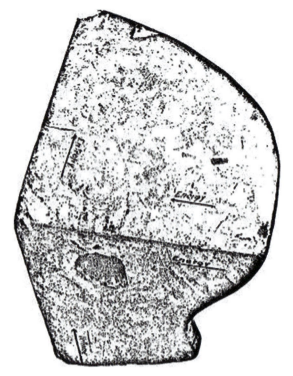

Figure 3.13 Another photo of a cast of the Gibbs surface.

The axes in this case are U, S and V. Thus, for example, the temperature and pressure can be obtained from the slope of the surface at each point.

Problem 3.1
Instead of using Eq. (3.1.1) we could have defined a temperature scale by choosing a convenient value of Boltzmann's constant. Let us choose the value $k_B = 1$. Since that will make k_B vanish from all our equations and we don't want to forget poor Boltzmann, let us call the result the Boltzmann scale. If we want to express the energy in ergs, what will be the temperatures of the triple point and the normal boiling point of water on the Boltzmann scale?

Problem 3.2
Fahrenheit invented a temperature scale in which zero was the lowest temperature reachable by mixing ice and salt, and 100 degrees was the average body temperature of his friends. In the resulting temperature scale, the triple point of water occurs at 32 °F and the NBP is 212 °F.

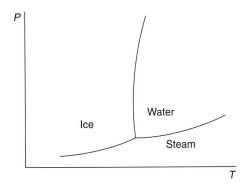

Figure 3.14 The phase diagram of water, showing the negative slope of the melting curve.

What temperature corresponds to zero kelvins? At what temperature do the Fahrenheit and Celsius scales have the same value? If we define another scale in which zero is the absolute zero of temperature (i.e. the same as zero kelvin), but the unit this time is the same as 1 °F, what is Boltzmann's constant? This scale is called the Rankine scale.

Problem 3.3
Is the isothermal compressibility of a real substance ever equal either to zero or to infinity? If the answer to either question is yes, use a sketch of a P–V–T diagram to help describe the circumstances.

Problem 3.4
The P–T–V diagram of water differs from the kind we described in that the melting curve has a negative slope at the triple point (Fig. 3.14). Explain why it is a necessary consequence of this feature that water is denser than ice when they are in equilibrium, and conversely why it is equally necessary that solid is denser than liquid in the case discussed in the text.

Problem 3.5
Helium is different from all other substances in that it never solidifies in contact with its own vapor, as shown in the P–T diagram in Fig. 3.15. Sketch the P–V phase diagram, including isotherms.

3.2 Ideal and non-ideal gases

Although atoms and molecules are electrically neutral, they always have electrical forces acting between them. The electrical forces come about

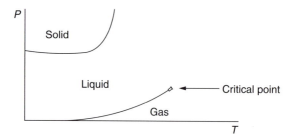

Figure 3.15 The phase diagram of helium.

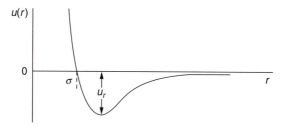

Figure 3.16 The pair potential between simple atoms.

either because their electrical charges are not symmetrically distributed
to begin with, or because each atom tends to push around the charges
on the others. The resulting forces between atoms can be represented by
a potential energy of interaction, or more simply a ***pair potential***, $u(r)$,
where r is the distance between the centers of a pair of atoms. The typical
form of $u(r)$ is sketched in Fig. 3.16.

The force at each point is actually the negative of the derivative or slope
of this curve. At large r the force is weakly attractive; at small distances
it is powerfully repulsive. In between there is a minimum in the potential
at which point the forces are balanced. The repulsive force at small
distances means the atoms are behaving as if they had hard cores. Very
roughly, when the atoms are the distance σ apart, we can think of their
hard cores as touching. Thus σ is twice the hard-core radius, or in other
words the hard-core diameter. The minimum potential energy of the
system is $-u_0$. u_0 is called the ***well depth***.

All the complicated phenomena we saw in the previous section may be
attributed to these forces between atoms. The connection shows up, for
example, in the fact that the critical temperature for each substance, T_C, is
roughly of order u_0/k for that substance, and the density of the condensed
material (liquid or solid) near the triple point is roughly $N/V \approx 1/\sigma^3$.

Table 3.2 *Properties of monatomic gases*

$PV = Nk_BT$ (equation of state)	$K_T = 1/P$
$U = \dfrac{3}{2}Nk_BT$	$\beta = 1/T$
$C_V = \dfrac{3}{2}Nk_B$	$S(T,V) = \dfrac{3}{2}\log(Nk_BT) - Nk_B \log\left(\dfrac{N}{V}\right) + \dfrac{3}{2}Ns_0$
$C_P = \dfrac{5}{2}Nk_B$	$S(T,P) = \dfrac{5}{2}Nk_B \log(k_BT) - Nk_B \log P + \dfrac{3}{2}Ns_0$
$\gamma = 5/3$	$s_0 = k_B\left[\dfrac{5}{3} + \log\left(\dfrac{mk_B}{2\pi\hbar^2}\right)\right]$

On the other hand, if the density of atoms is very small, in other words $N/V \ll 1/\sigma^3$, then the average potential energy acting between any pair of atoms will be small. If in addition the temperature is high, the kinetic energy of each atom will be much greater than its potential energy due to the other atoms. Under these conditions, we may ignore the potential energy acting between the atoms. The atoms then behave like the components of the perfect gas, whose quantum mechanical description we studied in Chapter 1. In precisely the same conditions of low density and high temperature, the thermodynamic description of the perfect gas reduces to that of the ideal gas, which we have used all along as a thermodynamic example (deducing the equations from the quantum mechanical description is a job we must postpone until we have mastered the machinery of statistical mechanics in Chapter 7). For reference, the properties of a monatomic ideal gas are listed in Table 3.2.

If the gas is made up of molecules rather than atoms, some of the formulas may change. The equation of state is not affected because that only involves counting translational states, but some of the energy of the system may go into tumbling motion or internal vibrations of the molecules (even atoms may have some internal complications). To take a particularly important example, the diatomic molecules O_2 and N_2 (which between them constitute nearly all of the air we breathe) at around room temperature have just about $(1/2)k_BT$ per atom of energy in internal vibration in addition to the $(3/2)k_BT$ per atom of kinetic energy that all atoms or molecules have when in the ideal gas state.

Problem 3.6

For air, $PV = Nk_BT$ to a very good approximation, and $U = (5/2)k_BT$ as just explained. Find C_P and γ.

In the P–V plane, the ideal gas is found at $V \gg N\sigma^3$, where σ^3 is roughly the volume per atom. In that region, the P–V isotherms are hyperbolic, $PV = $ constant. In the rest of the plane, the pair potential leads to departures from ideal gas behavior (see Fig. 3.4). At very high density, all substances solidify. At intermediate densities, there are wiggles in the P–V isotherms for $T > T_C$, and condensation into a liquid for $T < T_C$.

It is often convenient to have an equation to represent the behavior of gases and fluids when the ideal gas equation is not accurate enough. Many equations of state have been proposed. Two are particularly important.

The first of these is called the ***virial equation of state***. It is constructed by assuming that the ideal gas equation is a low-density approximation to $P(T, V)$. The higher density value of the pressure is corrected by adding terms of successively higher powers of N/V. The equation may be written

$$P = \frac{Nk_BT}{V}\left[1 + B(T)\frac{N}{V} + C(T)\left(\frac{N}{V}\right)^2 + \cdots\right] \qquad (3.2.1)$$

where $B(T)$ is called the ***second virial coefficient***, $C(T)$ is called the ***third virial coefficient***, and so on. B, C and the rest are functions only of the temperature.

Later on, in Chapter 8, we will see how $B(T)$ may be derived from the pair potential, $u(r)$. For now, let us just try to understand the general tendency of the first correction term. As we move to the right from the critical value of V in Fig. 3.4, we enter a zone where the gas is accurately represented by

$$P = \frac{Nk_BT}{V}\left(1 + B\frac{N}{V}\right)$$

where $B(N/V)$ represents the first effects of the potential energy of interaction between the atoms. We ask, will P be greater or less than that of an ideal gas? In other words, will B be positive or negative? The answer is not obvious because, as we have seen, $u(r)$ has both attractive and repulsive parts. If, on the average, the net effect is attractive, the gas will tend to pull itself together a bit, relative to the ideal gas. That means that P will be lower than ideal, so B must be negative. Conversely, a net repulsive effect will make B positive. The question of which of these effects dominates can be settled by comparing each with the ideal gas part of the energy, which is purely kinetic.

The energy of the ideal gas is $(3/2)Nk_BT$, so the average kinetic energy per atom is $(3/2)k_BT$, changing to $(5/2)k_BT$ for a diatomic gas, a change that will not affect this argument. The attractive part of $u(r)$ never has a magnitude larger than the well depth, u_0. Thus if $k_B T \gg u_0$ the attractive part of the potential energy between two atoms is unimportant compared with their kinetic energies, no matter what their separation. On the other hand, the repulsive part of the potential energy can be arbitrarily large if the atoms try to get too close together; the atoms are excluded from each other's hard cores. Therefore the first departure from ideality as the density increases at high temperature is an increase in the pressure. You can think of it as being due to the fact that the atoms find they have less than the full volume of the box to fly around in. In any case $B(T)$ is positive at high T. At lower T, the attractive part of the potential starts to become important, reducing B and eventually driving it to become negative. $B(T)$ is positive at high temperature and negative at low temperature, passing through zero in between.

The second important equation of state is called the ***van der Waals equation***. It is not particularly accurate quantitatively, but its qualitative behavior is very significant. With a little help, it predicts that gases condense into liquids along a coexistence curve that ends in a critical point. It had a dramatic impact when it was first introduced, in the Ph.D. thesis of Johannes Diderik van der Waals (1837–1923), in 1873. Even today it forms the basis of our understanding of these phenomena.

Johannes Diderik van der Waals

Surprisingly, for so fundamental an equation, it cannot be derived at all, but it can be arrived at with some simple, rather crude arguments similar to those we applied to $B(T)$ above. We start from the ideal gas equation, anticipating that we are going to argue that the potential energy of interaction modifies both the pressure and volume from their ideal values. Thus we write

$$P_I V_I = N k_B T \tag{3.2.2}$$

where the subscript I stands for ideal. We will replace the subscripted factors. The ideal volume is the volume of empty space each atom has to fly around in. It is less than the total volume because each atom is excluded from the volume occupied by the hard cores of the other atoms. Let us call the volume occupied by the hard core of a single atom b. Then the total excluded volume is Nb, and we have

$$V = V_I + Nb \tag{3.2.3}$$

We now assume that this substitution takes care of the effects of the repulsive part of the interatomic potential, and P_I needs to be modified only to take account of the attractive part.

In the 1860s J. C. Maxwell (the one who made plaster casts) tried to deduce the force law between neutral atoms at large distances, using measurements of the viscosity of gases that he made himself. He concluded that the potential went as r^{-5} so that the force went as r^{-6}. Today we believe that the potential goes as r^{-6} and the force as r^{-7}, but, of course, van der Waals could not have known that. The average attractive force between the atoms shows up directly in the pressure, which is reduced relative to the ideal pressure by an amount proportional to the sixth power of the mean distance between atoms, or in other words to the square of the mean volume per atom, V/N:

$$P = P_I - a(N/V)^2 \tag{3.2.4}$$

On solving Eqs. (3.2.3) and (3.2.4) for P_I and V_I and substituting into Eq. (3.2.2), we get

$$[P + a(N/V)^2](V - Nb) = N k_B T \tag{3.2.5}$$

This is the van der Waals equation of state.

As we have repeatedly stressed, any of our variables is fixed for a given sample by the values of any two others. Equations of state establish the relationship among P, V and T. That might not be the most useful possible combination. It might have been better to have the relationship among U, S and V, from which, as we have seen, everything else can be deduced. The information contained in the equation of state is less complete. For example there is no way to find the entropy or heat capacity of a system from its equation of state alone. Nevertheless, equations of state are useful because they relate the most easily measured variables, and therefore predict the most commonly observed parts of the system's behavior.

Problem 3.7
Find K_T and β for a gas obeying $P = (Nk_BT/V)[1 + BN/V]$, where $B(T)$ is the second virial coefficient.

Problem 3.8
Find K_T and β for a gas obeying the van der Waals equation of state.

Problem 3.9
According to our arguments, the constants a and b of the van der Waals equation ought to be related to the parameters σ and u_0 of the pair potential.

(a) How is b related to σ? (Answer $b = (2/3)\pi\sigma^3$.)
(b) On dimensional grounds alone (just getting the units straight), how should a be related to u_0 and σ? (Answer $a = Cu_0\sigma^3$, where C is a dimensionless constant, which should be of order 1 (why?).)

Problem 3.10
Show that if $V \gg Nb$ the van der Waals equation reduces to $P = (Nk_BT/V)[1 + NB/V]$, with $B = b - a/(k_BT)$. (Hint: it helps to use the approximation $1/(1 - x) \approx 1 + x$ if $x \ll 1$. Does this result have the form B ought to have?)

Problem 3.11
The second virial coefficient for helium is found to be $B(T)/k_B = 0.251 - 5.1/T$, where k_B is Boltzmann's constant and the units of B/k_B are K/atm. Estimate the percentage error in N/V at the normal boiling point if we calculate it using the ideal gas equation of state. The NBP of helium is 4.2 K.

Problem 3.12
Use the formula in Problem 3.11 and the result of Problem 3.9(a) to estimate the diameter of a helium atom.

Problem 3.13
When $V \gg Nb$ (as in Problems 3.10 and 3.11), we can estimate the correction to the ideal gas entropy by calling the volume in $S(T, V)$ the

ideal gas volume, V_I, then replacing it by $V - Nb$ as in Eq. (3.2.3). In this approximation, find C_V and C_P for a monatomic gas. Show that U (T, V) can be written $U = (3/2)Nk_BT - aN^2/V$.

3.3 The atmosphere

In this section we will apply what we've learned to the most familiar gas of all, the air we live in and breathe. The atmosphere is not a system in thermodynamic equilibrium. There are obvious variations from season to season and day to day, which we call the weather. However, those variations can be thought of as departures from an average, stable state. The pressure of the atmosphere varies with height because of gravity. Basically, the air at each level must support the weight of the air above it. We can analyze this situation by imagining a vertical column of air of unit area (1 square cm). Consider a thin slice of the column of thickness dh. The pressure on the top of the slice is equal to the force on the top, which is just Mg, where g is the acceleration of gravity and M is the total mass of all the air in the column above the slice. The force on the bottom of the slice is greater than that at the top because of the additional mass, ΔM, of the air inside the slice:

$$dP = -(\Delta M)g$$

where dP is the change in pressure across the slice (the minus sign is there because the pressure goes down as we go up). But $\Delta M =$ (mass per unit volume) × (volume of the slice), or

$$\Delta M = (mN/V)dh$$

(the volume is dh because the area is 1). Air is very nearly an ideal gas, so we can use $N/V = P/(k_BT)$:

$$dP = -(\Delta M)g = -mg(N/V)dh = [mg/(k_BT)]P \, dh$$

Here m is the average mass of an air molecule, $m = 28.8$ (times the mass of a proton). To find the change in pressure on going from $h = 0$ to some finite value of h, we simply integrate this equation with $mg/(k_BT)$ constant:

$$\frac{mg}{k_BT}\int_0^h dh = -\int_0^h \frac{dP}{P}$$

$$\frac{mg}{k_BT}h = -\log\left(\frac{P(h)}{P(0)}\right)$$

or, finally,

$$P(h) = P(0)e^{-mgh/(k_B T)} \qquad (3.3.1)$$

This is known as the barometric formula.

Problem 3.14
Where on Earth is the pressure of the atmosphere equal to 1 atm?

Problem 3.15
Using the barometric formula and the phase diagram of water, explain why people who live at higher altitudes almost always own a pressure cooker.

Problem 3.16
A barometer is a device that measures atmospheric pressure. How would you use a barometer to determine the height of a building? Suppose you have a mercury barometer that you can read to the nearest 0.5 mm Hg. How high does a building have to be before you can measure its height with an accuracy of 5%?

If the air were undisturbed, it would be at uniform temperature even though its pressure is not uniform. However, because of convection currents – updrafts and downdrafts – there is actually a gradient in the temperature of the air. This occurs because air is a poor conductor of heat. As the air moves up or down through the atmosphere, it is being expanded or compressed according to the pressure dependence on height as we have just studied. Since the air conducts little heat, the expansion or compression takes place almost adiabatically.

We can estimate the temperature of the air at each altitude by computing the temperature a bubble of air would have as it floats adiabatically upward through the atmosphere. The region of the lower atmosphere to which this estimate is applicable is called the ***troposphere***. Starting from the middle of the previous derivation, we have $dP/P = [mg/(k_B T)]dh$. As the bubble floats upward, it follows the adiabatic condition, $PV^\gamma =$ constant. Using the ideal gas equation to eliminate V for a given N, we find $P^{1-\gamma}T^\gamma =$ another constant. In differential form this becomes

$$\frac{dP}{P} = \frac{\gamma}{\gamma - 1} \frac{dT}{T}$$

By substituting for dP/P in the equation we started with, we find for the temperature gradient

$$\frac{dT}{dh} = -\frac{\gamma - 1}{\gamma}\frac{mg}{k} \qquad (3.3.2)$$

For air, $\gamma = 7/5$, so $\gamma - 1 = 2/7$. If we multiply both m and k by Avogadro's number, N_0, we get $mN_0 \simeq 29$ for air and $kN_0 = R = 8.2 \times 10^7$ erg/mole degree (R is called the gas constant) and $g = 980$ cm/s^2. We get

$$\frac{dT}{dh} = -\frac{2}{7} \times \frac{29 \times 980}{8.2 \times 10^7} \simeq -1 \times 10^{-4}\ {}^\circ\text{C/cm} = -10\ {}^\circ\text{C/km}$$

In other words, air in the mountains at 1000 meters ought to be about $10\ {}^\circ$C cooler than air at sea level. There are two principal reasons why the temperature of the atmosphere usually falls less rapidly than this estimate with increasing altitude. The first of these, at lower levels, is the effect of water vapor, and the second, at higher levels, results from the ozone layer in the atmosphere.

We have calculated the cooling due to adiabatic expansion of dry air. If the relative humidity at sea level is not zero, the water vapor in the atmosphere also expands adiabatically and cools. As it does so, it tends to condense, liberating heat, which has the effect of decreasing the temperature gradient. This is also the reason why clouds form. At higher elevations, there is a layer of ozone, which efficiently intercepts ultraviolet radiation from the Sun and turns it into heat, warming the upper atmosphere. Thus, as one ascends from the ground, the temperature first decreases, then increases somewhat.

Problem 3.17
We derived the barometric formula assuming the temperature of the atmosphere to be uniform. Derive a more accurate formula taking into account the adiabatic expansion of dry air. Estimate the height up to which the original formula is a good approximation.

4

The laws of thermodynamics

4.1 The origin and meaning of the first and second laws

Of all the tales in the repertory of human folklore, one of the best may be one that is hardly ever told. It is the epic saga of the second law of thermodynamics. The second law was born in an attempt to improve the steam engine, and it went on to predict the ultimate fate of the Universe. It captured the popular imagination, defined the meaning of time, and led to the discovery of quantum mechanics. It was born, like some mythical beast, before its putative mother, the first law of thermodynamics.

Before we get swept away by the mighty tide of these powerful laws, let us remind ourselves that we already know everything there is to know about what they say and why they are true. The first law expresses, in a special way, the conservation of energy. We examined it at some length earlier. The second law says essentially that the entropy of an isolated system increases until it can get no larger. We've already seen that that's true. What we have to say here is therefore mere elaboration.

Before the middle of the nineteenth century, it was believed that heat was a fluid, called caloric, which could flow from one body to another. That view differed from our present understanding in that each body was thought to possess a definite amount of caloric, which, being a fluid, could neither be created nor be destroyed. It was a successful theory, but it wasn't quite right.

It is hard for us today to imagine a way of thinking in which work could not be turned into heat, or heat into work. Could it be that, before the scientists showed how in the middle of the nineteenth century, nobody ever thought of rubbing his or her hands together on a cold day? More to the point, by the 1820s rails had begun to slice through the land, and locomotives were shattering the peace of the countryside. What were locomotives if not visible, palpable evidence that heat obtained from burning coal could be turned into work?

Nicolas Léonard Sadi Carnot

It is possible that had the first law been known in the 1820s, the second law would have been much more difficult to discover. The essential step that led to the second law was taken by a young French military engineer named Nicolas Léonard Sadi Carnot (1796–1832). Carnot set out, as had many before him, to improve the efficiency of the steam engine. His approach differed from his predecessors, however, in that, instead of looking for a practical way to make a better steam engine, he asked a theoretical question. What, he asked, is the most efficient steam engine possible? He tried to picture an ideal engine that would illustrate the intrinsic limitations on the very physical processes that made an engine work.

If Carnot had known about the first law, he might have been tempted to find a way of turning all of the heat from the fuel into work. Instead, the picture he developed was an analogy to a water wheel. Water falls from a high level, makes the wheel turn doing work, and winds up at a lower level. The water itself is not consumed. In the same way, Carnot pictured caloric starting at high temperature, causing work to be done while falling to low temperature, but not getting used up. It was natural to think, then, of heat being more useful at high temperature than at low temperature. That idea leads directly to the second law of thermodynamics.

We can see the connection, in the language of physics, by considering what happens when a small quantity of heat, ΔU, is conducted from a warm body at temperature T_H to a cool body at temperature T_L. The entropy of the warm body is reduced by $-\Delta S_H = -\Delta U/T_H$ while that of the cool body is increased by $\Delta S_L = -\Delta U/T_L$. The net change in entropy is

$$\Delta S_L - \Delta S_H = \Delta U \left(\frac{1}{T_L} - \frac{1}{T_H} \right)$$

which is positive. This increase in the entropy of the system does not surprise us: The system is tending toward equilibrium. The heat will never spontaneously return to the hotter body simply because that is a much less probable state of the combined system.

Carnot was the first to think of analyzing an engine by following a fixed quantity of its working fluid through a complete cycle. (We still do the same today.) He then envisioned an abstract, idealized cycle that could be run *reversibly*. In other words, a quantity of heat, falling from a high temperature to a lower one, could produce a certain amount of work. The same quantity of work, running the same cycle in reverse, would lift the same quantity of heat back up to the high temperature. Carnot's starting point, aside from the caloric theory, was the sound engineering principle that you could not build a perpetual motion machine. He showed (we will see his proof later) that the principle would be violated if there existed any machine more efficient than his ideal cycle. He had therefore reached the intrinsic limit, designed the most perfect engine.

Carnot's analysis did not lead to any improvement in the steam engine. Nevertheless, on the basis of his single publication, he was in his day a well-known and widely respected steam engineer. For a long time after his death, however, physicists ignored his ideas, which were to have the most profound impact imaginable on physics. Most remarkable of all is that the structure he created survived the discovery of the first law.

If ever there was an idea whose time had come, it was the first law of thermodynamics, in the middle of the nineteenth century. In fact, it was discovered at least nine separate times.

Credit for a scientific discovery sometimes goes not to the first discoverer but to the last. The last discoverer of the first law was James Prescott Joule (1818–1889), the son of a retired brewmaster, who did his experiments in a private laboratory built for him by his father. In 1849 Joule's masterly determination of the amount of work needed to produce a given amount of heat established forever the interconvertability of heat and work.

James Prescott Joule

Lord Kelvin

While Joule was busy in his laboratory, the theoretical physicists discovered Carnot. In 1848, William Thomson (1824–1907), later to become Lord Kelvin, put Carnot's verbal and geometric arguments into mathematical form. Two years later, Rudolf Clausius (1822–1888) reformulated

Rudolf Clausius

them to be consistent with the first law and invented the idea of entropy. He combined the two laws into a maxim that was later adopted by the great American thermodynamicist J. W. Gibbs:

> The energy of the Universe is conserved. The entropy of the Universe tends to a maximum.

Carnot, the engineer, had thought of a way to analyze steam engines. Clausius, the physicist, was inclined to generalize a bit, to include the Universe.

4.2 Ideal engines

> There was a young man named Carnot
> Whose logic was able to show
> For a work source proficient
> There's none so efficient
> As an engine that simply won't go
> *Written by the author as an undergraduate, Brooklyn College, 1960*

Our purpose in this section is to do what Carnot did: not to design a real working engine, but rather to extract the essence of what we shall call "engineness". We shall make use of some advances in knowledge that Carnot didn't have. All heat engines must extract heat from some source, do work such as lifting a weight or driving a car, then somehow return to their initial state so they can do it again. We can start our analysis at the first step, extracting heat.

Let us borrow from Chapter 1 a box of ideal gas to serve as the source of heat. It has energy U (some of which will be extracted) divided among N particles in volume V with consequent entropy S. To this we attach a machine of unspecified design, which will use some of the energy. What we know about the machine is that, like everything else, it obeys the laws of quantum mechanics. The machine has some number of possible microscopic states, which will increase if energy flows into it from the box. At the same time the number of microscopic states of the box will decrease. We also know that energy is likely to flow from the box to the machine only if that makes the product of those two numbers bigger. In other words, if energy does flow, it is because that process leads to a more probable state of the combined system, and the energy will therefore not return to the box.

We can translate the argument into the usual symbols. The box has some number of microscopic states, Γ_{box}, which is a monotonic function of its energy, U. The machine also has some number of microscopic states, $\Gamma_{machine}$. Energy will flow from the box to the machine only if it increases the quantity $S_{total} = k_B \log(\Gamma_{box}\Gamma_{machine})$, where S_{total} is the entropy of the entire system.

Generalizing this argument, we can formulate our own homegrown version of the second law:

> Nature will not do anything unless we entice her into it with the promise of higher entropy.

Notice that the entropy of the box alone decreases in the process we described. It is the entropy of the combined system that must increase. It is a bit pretentious, although correct, to refer to this as an increase in the entropy of the Universe. The fact that the entropy of the Universe increases in any real event is difficult to make use of since we can't say how much it must increase. All we can say with assurance is that the entropy of the Universe cannot decrease in any real event. We are thus free to imagine events in which the increase in entropy of the Universe is arbitrarily small; in fact we take the increase to be zero. Such an event is said to be *reversible*. Reversible processes, which are not in fact possible, are nevertheless the fundamental building blocks of thermodynamic analysis.

Let us drive the point home:

> A reversible process is one in which the entropy of the Universe does not change.

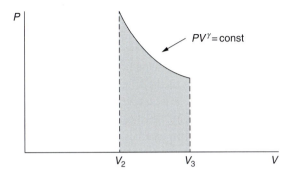

Figure 4.1 An adiabatic expansion.

Our basic purpose is to construct a reversible engine. It must extract heat, do work, and return to its initial state so it can start again. Moreover, every single step must be reversible since, once the entropy of the Universe has increased, it can never come back down again. We can begin by reviewing processes that we know to be reversible.

One such process is an *adiabatic expansion* of an ideal gas. We can imagine the gas in a cylinder with a piston, isolated from the rest of the Universe so that no heat flows in or out. In these conditions the gas expands, pushing out the piston. The entropy of the rest of the Universe does not change, but the volume of our gas increases, work is done, and the energy and temperature of the gas are reduced.

Let us visualize the process on a plot of P versus V for the gas in the cylinder. As we saw earlier, the gas follows the curve

$$PV^\gamma = \text{constant} \tag{4.2.1}$$

Thus an adiabatic expansion from V_2 to V_3 would look like Fig. 4.1.

The work done by the gas is

$$W_{23} = \int_{V_2}^{V_3} P\, dV$$

On the graph in Fig. 4.1, the work is the area under the curve, which is shaded.

Having arrived at V_3 and therefore some pressure and temperature, what can we do next without increasing the entropy of the Universe? We could compress adiabatically back to V_2, but that would be useless. We must instead find some other reversible path to follow.

It turns out that another reversible step we can take is an *isothermal* expansion or compression. In an isothermal step, the entropy of the gas inside the cylinder does change, but the entropy of the Universe does not. To see why not, consider the following system.

Let our cylinder be attached to a very large box of ideal gas. Now let the gas in the cylinder expand, say from volume V_1 to a larger volume, V_2. We can think of this as a sort of modified adiabatic expansion, since the system of cylinder plus box is isolated from the rest of the Universe. Thus, from the point of view of the combined system, the process is adiabatic, but only a small part of it expands. The amount of work done is

$$W_{12} = \int_{V_1}^{V_2} P \, dV$$

Energy is conserved, $\Delta(U_{\text{box}} + U) = -W_{12}$, where U is the energy of the gas in the cylinder, and U_{box} that of the gas in the box. If for simplicity we take both gases to be monatomic, we have

$$U = \frac{3}{2} N k_{\text{B}} T$$

$$U_{\text{box}} = \frac{3}{2} N_{\text{box}} k_{\text{B}} T$$

so that, when the work is done, the whole system cools down by

$$\Delta T = -\frac{W_{12}}{(3/2)(N_{\text{box}} + N) k_{\text{B}}}$$

There is no limit to how big the box can be, so we take it (and the amount of gas inside) to be infinite. Then $N_{\text{box}} = \infty$ and $\Delta T = 0$. The process is isothermal. A body from which we can extract energy without changing its temperature is just what we called a temperature bath. It is, like the reversible process itself, another one of those handy thermodynamic idealizations we use.

Now consider what happens in this isothermal expansion. The cylinder goes from volume V_1 to volume V_2 at constant temperature. Work W_{12} is done, but the energy of the gas in the cylinder, $U = (3/2)N k_{\text{B}} T$, doesn't change. The energy must therefore have been extracted entirely from the bath. The net effect of an isothermal expansion is to convert heat from the bath into work.

We now have at hand all of the elements necessary to construct a complete reversible cycle, the same one constructed by Carnot. The first two steps are as follows.

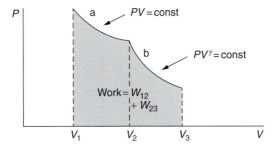

Figure 4.2 An isothermal step plus an adiabatic one.

(1) Starting at V_1 and temperature T_H, attach the cylinder to a bath at T_H and expand isothermally to V_2. Work W_{12} is done.
(2) Disconnect the cylinder and expand further, this time adiabatically, to volume V_3. The temperature drops to some new value, T_L. More work, W_{23}, is done. The two steps are shown in the combined graph of Fig. 4.2.

The gas in the cylinder, which is our working fluid, started at volume V_1 and temperature T_H. Its entropy then was, say, S_1. Now it has larger volume, V_3, lower temperature, T_L, and higher entropy, say, S_2. It extracted from the bath at T_H an amount of heat given by

$$Q_H = \int_{S_1}^{S_2} T\, dS = T_H(S_2 - S_1) \tag{4.2.2}$$

It has performed an amount of work $W_{12} + W_{23}$.

> **Problem 4.1**
> If our working fluid consisted of N atoms of an ideal monatomic gas, find the heat extracted and work done up to this point in terms of N, V_1, V_2, T_H and T_L.

The rest of our job is to return the working fluid to its initial state by reversible steps, doing less work than we have already extracted so that the entire cycle produces net, usable work. We must somehow reduce the entropy from S_2 back to S_1, and reduce the volume from V_3 to V_1.

> **Problem 4.2**
> If we get back to V_1 and S_1, we don't have to worry about the temperature. It will be T_H. How do we know that?

Obviously it does no good to compress the gas while its entropy remains at S_2. That just takes us back along the same adiabatic path, requiring that we put back as much work as we got out on the second step. Instead we

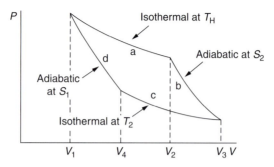

Figure 4.3 The complete Carnot cycle.

must first reduce the entropy. In other words, to get back to the initial state while still getting work out, *we must extract at low temperature some of the work we put in at high temperature.* That requirement applies to all heat engines and is the reason why there is a thermodynamic limit to their efficiency. It is also the reason why your car has a radiator.

To complete the cycle, we need a second temperature bath, this one at T_L. Then we complete the cycle with the following steps.

(3) Attach the cylinder to the bath at T_L and compress until the entropy reaches S_1.
(4) Remove the cylinder and compress again, this time adiabatically, until the volume reaches V_1.

The whole cycle is shown in Fig. 4.3.

The work put back in the last two steps is the area under the corresponding curves. On the whole cycle, a net amount of work equal to the area inside the closed path is done. We represent that work as

$$W = \oint P \, dV \tag{4.2.3}$$

Here the symbol $\oint P \, dV = \oint V \, dP$ means integrate around a closed path in the clockwise direction:

$$\oint P \, dV = \int_{V_1}^{V_2} P \, dV + \int_{V_2}^{V_3} P \, dV + \int_{V_3}^{V_4} P \, dV + \int_{V_4}^{V_1} P \, dV$$

Problem 4.3
Find V_4 in terms of N, V_1, V_2, T_H and T_L.

Problem 4.4
Compute $\oint P \, dV$ in terms of the same variables.

The heat we had to extract at T_L is given by

$$Q_L = -\int_{S_2}^{S_1} T \, dS = T_L(S_2 - S_1)$$

The difference between the heat in, Q_H, and the heat out, Q_L, must be the work

$$W = Q_H - Q_L = (T_H - T_L)(S_2 - S_1)$$

Notice that, since $Q_H = T_H(S_2 - S_1)$ and $Q_L = T_L(S_2 - S_1)$,

$$\frac{Q_H}{Q_L} = \frac{T_H}{T_L} \qquad (4.2.4)$$

This equation is used to define absolute temperature in formal thermodynamics.

The cycle we have constructed here is the Carnot cycle; it is the same as the one constructed by Carnot himself in 1824. We can see how he would have arrived at it with his water wheel analogy: All heat transfers take place at constant "height", that is, under conditions in which the temperature is reversible. At each isothermal step, heat can be transferred either way, from bath to engine or from engine to bath. The whole thing can equally well be run backwards, in which case it is a Carnot refrigerator instead of a Carnot engine. The refrigerator traverses the same path, two isotherms and two adiabatics, but in the counter-clockwise direction. The net work is negative: Work must be put in, having the effect of extracting heat Q_L at T_L and depositing heat $Q_H = W + Q_L$ at T_H.

To the engineer, the merit of an engine lies in getting as much work as possible out of an expenditure, Q_H, of heat from the fuel that must be burned. We therefore define the efficiency of the engine to be

$$\eta = \frac{W}{Q_H} = \frac{Q_H - Q_L}{Q_H} = 1 - \frac{Q_L}{Q_H}$$

We have already seen that, for a Carnot engine operating between baths at T_H and T_L, $Q_L/Q_H = T_L/T_H$, so that

$$\eta_{\text{Carnot}} = 1 - T_{\text{L}}/T_{\text{H}} \tag{4.2.5}$$

Thus even an ideal reversible engine would not be perfectly efficient.

Problem 4.7

A typical combustion temperature is 2000 K. Find the efficiency of a Carnot engine extracting heat from a source at 2000 K and rejecting waste heat into the atmosphere at 300 K.

We said in the last section that Carnot was able to show that his engine was the most efficient possible. His proof was as follows.

Suppose there was a super-machine capable of operating between temperatures T_{H} and T_{L} with efficiency η^* greater than the Carnot efficiency. That means that, if it extracts heat Q_{H}, it would produce work given by

$$\eta^* = W^*/Q_{\text{H}}$$

dumping heat $Q_{\text{L}}^* = Q_{\text{H}} - W^*$ at T_{L}. Now let us construct a Carnot engine operating between the same baths and extracting the same amount of heat, Q_{H}. It then does work given by

$$\eta_{\text{Carnot}} = W/Q_{\text{H}}$$

Since we have assumed that $\eta^* > \eta$ it follows that $W^* > W$.

Now we run our Carnot engine backwards, as a refrigerator (the fact that it is reversible is its key property). We now need work, W, to extract heat Q_{L} at T_{L}, dumping Q_{H} at T_{H}. In fact, we can use our super-machine to run the refrigerator, since it makes available more than enough work, W^*. The two machines running together have the effect of doing net work $W^* - W$, but one machine extracts heat from the bath at T_{H} while the other puts it right back. The whole thing works, therefore, without burning any fuel (it gets its energy, $Q_{\text{L}} - Q_{\text{L}}^*$, from the low-temperature bath, i.e. from the heat in the atmosphere). This is just the kind of perpetual motion machine that Carnot assumed to be impossible, so the super-machine could not exist. Carnot had successfully demonstrated that the Carnot cycle was the most efficient engine possible operating between any two temperatures.

Problem 4.8

One mole of monatomic ideal gas goes through a Carnot cycle in which it expands isothermally at temperature T_0 from volume V_0 to volume V_1. Later it's compressed isothermally at temperature T_2 from volume V_2 to volume V_3.

Find the volumes V_2 at the beginning of the isothermal compression and V_3 at the end. If we define the compression ratio r to be the volume V_2 after the adiabatic expansion to the volume V_1 before that step, write the efficiency in terms of r.

One reason why we cannot perform an ideal Carnot cycle is because we need temperature differences to make heat flow. Suppose we make our working fluid (still an ideal gas) go through the cycle described above, with heat exchanged on the high- and low-temperature steps at T_H and T_L. In both steps contact is made to the cylinder by way of thermal links with thermal resistance R (thermal resistance is to thermal conductivity exactly what electrical resistance is to electrical conductivity.)

Find the times t_0 and t_2 spent at each of the isothermal steps at temperatures T_0 and T_2, respectively. Assuming the adiabatic strokes are very fast compared with t_0 and t_2, find the power (work per unit time) of the machine in terms of r and the four temperatures T_0, T_2, T_H and T_L.

4.3 Tools of the trade

In this section we'll work out some of the tools we'll need to analyze engines and refrigerators.

All engines extract heat at high temperature, turn some of it into work, and reject the rest at low temperature. Any engine can be schematically represented as in Fig. 4.4, where the baths are at temperatures T_H and T_L and the Qs and Ws are positive, flowing in the directions shown by the arrows. A refrigerator looks like Fig. 4.5.

In Carnot's proof of the last section, we argued that there could be no super-machine more efficient than the Carnot cycle by combining the two machines as in Fig. 4.6. This diagram makes it immediately obvious that net work is being done without extracting any heat at T_H.

Another way to describe any engine is to say that it produces work while going through a complete cycle and returning to its initial state. If each step is reversible or can be idealized to a reversible process, the cycle can be represented on a P–V diagram for the working fluid. A typical engine might look something like Fig. 4.7.

If the cycle is followed in the direction shown, the area inside the closed loop gives the work done by the engine in each cycle, $\oint P \, dV$.

Every point on the P–V plane represents a possible equilibrium point of the working fluid of the engine. There are certain processes that could conceivably be used in a real cycle that cannot be represented in the plane

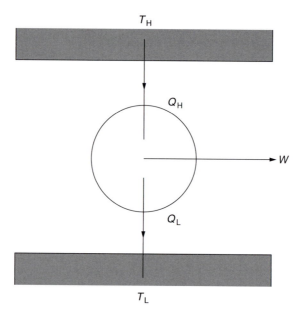

Figure 4.4 A heat engine.

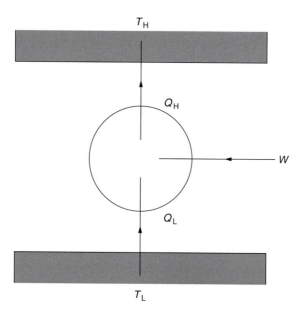

Figure 4.5 A refrigerator.

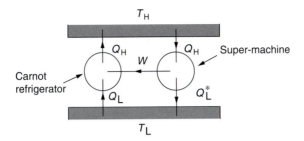

(Attention: This device is thermodynamically forbidden.)

Figure 4.6 A Carnot refrigerator attached to a super-machine.

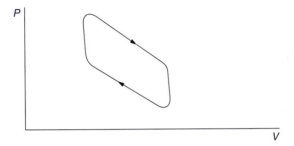

Figure 4.7 A typical engine.

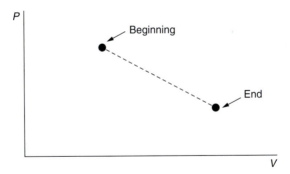

Figure 4.8 The free expansion of a gas.

because the fluid does not pass through equilibrium states. An example is a free expansion of a gas, as discussed earlier. We can find the beginning and end points in the plane, but in between, when the gas is expanding freely into a vacuum, it is not passing through states that can be represented in the plane. It looks something like Fig. 4.8. We cannot say in this case that the work done is $W = \int P\, dV$ because there is no particular path to integrate along.

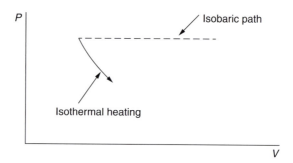

Figure 4.9 The isobaric heating of a gas.

Problem 4.9

We found earlier that in a free expansion of an ideal gas, the temperature remains constant. If Q is the amount of heat absorbed by the gas during the process, is it true that $Q = T \Delta S$, where ΔS, is the change of entropy of the gas? How much work is done in the free expansion? If ΔU is the change in energy of the gas, account for each term, including sign, in the first law, written in the form $\Delta U = Q - W$.

There are other kinds of processes that, although intrinsically irreversible, can nevertheless be idealized and represented in the plane. For example, in a gas turbine engine the gas is heated isobarically (at constant pressure) by burning fuel directly inside it. We cannot return the heat to the combustion products, thereby reconstituting the fuel; the process is quite irreversible. We can, however, imagine representing the isobaric heating of a gas on the $P-V$ plane. As T increases at constant P, V will increase, so it looks like Fig. 4.9.

The work done by the gas in this step is

$$W = \int P\, dV = P(V_{\text{final}} - V_{\text{initial}})$$

We just pretend that the heat came from outside somehow (it can't make any difference thermodynamically where the fuel was burned) and represent the process as a reversible step.

There is still a conceptual problem, however. The only reversible steps we know how to perform are isothermal and adiabatic. How do we imagine a reversible, isobaric heating? Heat, as we have seen, can be passed reversibly from one body to another only if the bodies are at the same temperature. In an isobaric heating, the temperature of the gas is continually changing. To do the job, we need a series of temperature baths differing by small steps in temperature. First we heat the gas

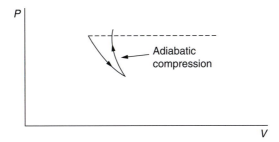

Figure 4.10 Two different paths.

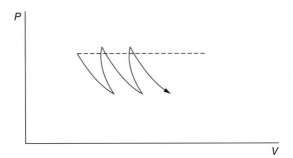

Figure 4.11 Recovery from two different paths.

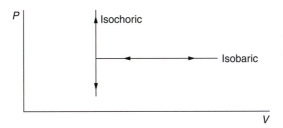

Figure 4.12 Isochoric and isobaric steps.

isothermally at the first bath. This causes V to increase but P drops, see Fig. 4.10. Then we bring P back by an adiabatic compression, increasing T to a new, higher value (Fig. 4.11). Then we compress adiabatically again (Fig. 4.12) and so on. Taking smaller and smaller steps, we produce as nearly as possible an isobaric step.

The point, then, is this: Any path that can be represented on a $P-V$ graph is, in principle, reversible. The Carnot cycle is special because it is the only reversible cycle that uses only two temperature baths. Many other reversible paths are possible if we use more temperature baths.

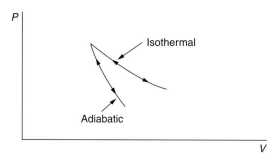

Figure 4.13 Isothermal and adiabatic steps.

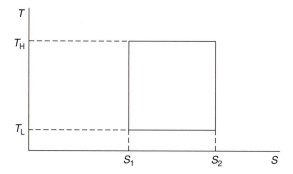

Figure 4.14 The Carnot cycle on a T–S diagram.

To sketch cycles in the P–V plane, we must have some idea of what various possible steps look like. For example, an isobaric step is horizontal and an isochoric step is vertical (Fig. 4.12).

An isothermal step has a slope given by

$$\left(\frac{\partial P}{\partial V}\right)_T = -\frac{1}{VK_T}$$

Since K_T is always positive, an isothermal step always has a negative slope. The slope of an adiabatic step is

$$\left(\frac{\partial P}{\partial V}\right)_S = -\frac{1}{VK_S}$$

But $K_T \geq K_S \geq 0$, so an adiabatic step also has a negative slope, but it is steeper than an isothermal slope (Fig. 4.13).

Problem 4.10

Show that these statements about adiabatics and isothermals are true for an ideal gas. Are they ever true for a van der Waals gas? Careful, this is tricky!

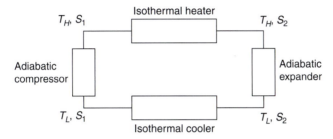

Figure 4.15 Various steps.

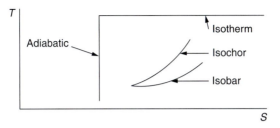

Figure 4.16 Another way to represent a Carnot cycle.

To analyze any engine cycle, a $P-V$ diagram is a useful place to start, since it shows graphically the work being done. There is an obviously complementary diagram that shows the heat used in the same way: a $T-S$ diagram. On a $T-S$ diagram, the isotherms and adiabats have obvious shapes. In fact, the Carnot cycle, which uses only those steps, has the simple form shown in Fig. 4.14. It's easy to see that the net heat used is the area inside the box, $\oint T\, dS$.

An isobaric step on a $T-S$ diagram has the slope

$$\left(\frac{\partial T}{\partial S}\right)_P = \frac{T}{C_P}$$

which is positive, and, for a gas at least, increases with increasing T. For an isochoric step,

$$\left(\frac{\partial T}{\partial S}\right)_V = \frac{T}{C_V}$$

which is also positive and generally larger than the isobaric slope because $C_P > C_V$ (Fig. 4.15).

Problem 4.11
Consider a cycle, which is a rectangle in the $P-V$ plane, with pressures and volumes P_1, P_2, V_1 and V_2 for one mole of an ideal gas working

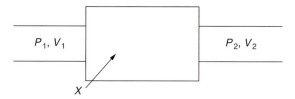

Figure 4.17 One element of an engine.

fluid. Sketch the $P-V$ and $T-S$ diagrams. Find T at each corner, the heat in at that point and work done at each step, and the efficiency of the cycle. Is the engine reversible? If it uses only one high-temperature and one low-temperature bath, how does its efficiency compare with that of a Carnot cycle?

In many cases the working fluid of an engine passes through a series of engine components in which heat or work is exchanged, changing its pressure, volume and temperature. The fluid can be imagined to flow to the next component in a new thermodynamic state. For example, we can imagine a Carnot engine running as a continuous flow cycle as shown in Fig. 4.16.

For the Carnot ideal gas cycle, of course, we already know exactly what is supposed to happen inside of each component. For real engines, however, we need some way to analyze the change in the thermodynamic state of the working fluids at each point in the cycle.

Suppose the fluid flows through a pipe at pressure P_1, occupying a volume V_1 per mole, into a certain component. Inside the component, heat may be added or extracted, and work may be done on the fluid or by it. Moreover, the process may be reversible in principle or intrinsically irreversible, like a free expansion. The fluid then emerges into another pipe at P_2, each mole now occupying V_2. Schematically, we have the situation shown in Fig. 4.17, where X may be Q or W, in or out. Can anything at all be said about so general a case?

We cannot assume that the fluid passes through possible equilibrium states inside the component. That's another way of saying we can't assume the process is reversible. However, it is in equilibrium before it enters and after it has exited the component, and we can keep track of all energies in the problem. The energy (per mole, say) at the end U_2 will be equal to the energy on the way in, U_1, plus the sum of all changes that take place along the way. The following things happen.

1. Work is done on the fluid on its way into the component. The next slug of fluid acts like a piston, pushing the mole we are watching into the

component at constant pressure. The volume of the fluid in the inlet pipe goes from V_1 to V_{int}. Since $\Delta U = Q - W$, the energy changes by $-W_1$, where

$$-W_1 = -\int_{V_1}^{V_{int}} P_1 \, dV = +P_1 V_1 - P_1 V_{int}$$

2. X is added inside.
3. The fluid does work on the way out, pushing out the previous slug at constant pressure, P_2, increasing its volume from V_{int} to V_2:

$$-W_2 = -\int_{V_{int}}^{V_2} P_2 \, dV = -P_2 V_2 + P_2 V_{int}$$

On adding up all the contributions, we have

$$U_2 = U_1 - W_1 + X - W_2 = U_1 + P_1 V_1 + X - P_2 V_2$$

or

$$X = (U_2 + P_2 V_2) - (U_1 + P_1 V_1)$$

Thus the heat or work exchanged in the component shows up as the change in the quantity $U + PV$ of the working fluid. The quantity $U + PV$ is, of course, the enthalpy of the fluid.

Example 4.3.1

Suppose the component is simply a fine nozzle, with the incoming gas, at P_1, squirting through to low pressure at P_2. What is X?

Solution.
No work is done, nor does heat go in or out: $H = H(S, P)$. This is the Joule–Thomson throttling process (see Problem 2.19). Enthalpy is conserved, $\Delta H = 0$. The process is irreversible.

Problem 4.12
What is X if the component performs a free expansion?

Recall that, since $H = U + PV$, we have $dH = T \, dS + P \, dV$, and S and P are the proper variables of H. Thus if $H = H(S, P)$ is known, all thermodynamic quantities can be deduced.

Problem 4.13
Construct $H(S, P)$ for the ideal gas, using Eq. (1.1.4).

Problem 4.14
Prove that $(\partial T/\partial P)_S = (\partial V/\partial S)_P$.

Enthalpy is a useful quantity to know. For example, if oil is burned in constant-pressure combustion, we get

$$\Delta H = \int T \, dS = Q$$

Thus it is enthalpy that is stored in fuel. By the same logic, if we wish to boil water, turning it into steam at constant (say, atmospheric) pressure, and therefore at constant temperature, the amount of heat we need, the ***latent heat of vaporization***, L, is

$$L = Q = \int T \, dS = T(S_g - S_1) = H_g - H_1$$

where the subscripts g and l stand for gas and liquid. Finally, to reiterate our point that prompted us to recall the idea of enthalpy, for any flow process (constant pressure in, another constant pressure out),

$$X = H_{\text{out}} - H_{\text{in}}$$

where X is the work or heat in or out in the process. For a given amount of ideal gas, we can easily find the enthalpy at any temperature and pressure. For real substances used in real engines, such as water and steam, enthalpies are carefully measured and tabulated.

Problem 4.15

One mole of fluid at temperature T_0 occupies volume V_0. It is made to undergo a small change in pressure, δP, by throttling. Find the change of entropy of the Universe.

5

The Boltzmann factor and
the partition function

5.1 The Boltzmann factor

A quantum mechanical sample (a single atom, a molecule or even a box of perfect gas) in thermal contact with a large body of matter (called a temperature bath) at temperature T has some probability of being found in any of its possible quantum states. The state might involve kinetic energy, rotational energy, internal vibrational energy, magnetic energy, electronic excited states and so on. Let us call any single microscopic state α, and label its energy ε_α. The problem before us is to answer this question: What is the probability of finding the sample in the state α?

To answer the question, we first imagine the sample and bath together to constitute a closed system, isolated from the rest of the Universe, with some total energy, U_0. In equilibrium the system has Γ_0 possible states, and entropy S_0,

$$S_0 = k_B \log \Gamma_0 \text{ (in equilibrium)} \qquad (5.1.1)$$

Γ_0 is the product of the number of possible states of the sample, Γ_{sample}, and the number of possible states of the bath, Γ_{bath}, when the system is in equilibrium with total energy U_0:

$$\Gamma_0 = \Gamma_{\text{sample}} \Gamma_{\text{bath}} \text{ (in equilibrium).} \qquad (5.1.2)$$

Now, instead of imagining the system to be in equilibrium, let us instead specify that the sample is in state α with energy ε_α. The bath is always internally in equilibrium. It has in this case energy $U_0 - \varepsilon_\alpha$, giving it some number of possible states, which we will call $\Gamma_{\text{bath}}^\alpha$, the number of possible states of the bath when the sample is in state α. Since the sample is in state α, its number of possible states is

$$\Gamma_{\text{sample}} = 1$$

so the number of possible states of the system is

$$\Gamma^\alpha_{\text{system}} = \Gamma_{\text{bath}} \Gamma_{\text{sample}}$$

$$= \Gamma^\alpha_{\text{bath}} \tag{5.1.3}$$

The probability of finding the sample in the state α is simply the number of states of the system in which the sample happens to be in the state α divided by the total number of states of the system in equilibrium. We will call the probability w_α (always a pure number smaller than one). Then,

$$w_\alpha = \Gamma^\alpha_{\text{system}}/\Gamma_0 = \Gamma^\alpha_{\text{bath}}/\Gamma_0 \tag{5.1.4}$$

The probability w_α plays a crucial role in statistical physics. Because the sample is always in some state, the probabilities of all the states must add up to 1:

$$\sum_{\text{all states}} w_\alpha = 1 \tag{5.1.5}$$

If a quantity, say, f_α, has a definite value in each quantum state of the sample, then its average or thermodynamic value is given by

$$\bar{f} = \sum_\alpha f_\alpha w_\alpha \tag{5.1.6}$$

This formula is not as unfamiliar as it may seem at first sight. It is exactly the same as the formula

$$\text{Average score} = \sum_{\text{all possible scores}} (\text{score})(\text{fraction of students who got this score})$$

In particular, for the thermodynamic case, f might be the energy of the sample. Then the thermodynamic or average energy is

$$\bar{\varepsilon} = \sum_\alpha \varepsilon_\alpha w_\alpha \tag{5.1.7}$$

The purpose of this section is to find a general expression for w_α.

When the sample is in the definite state α, the bath has entropy

$$S^\alpha_{\text{bath}} = k_B \log \Gamma^\alpha_{\text{bath}} \tag{5.1.8}$$

On subtracting this entropy from the equilibrium entropy of the system, Eq. (5.1.1), we get

$$S_0 - S^\alpha_{\text{bath}} = -k_B \log(\Gamma^\alpha_{\text{bath}}/\Gamma_0)$$

$$= -k_B \log w_\alpha \tag{5.1.9}$$

(we have used Eq. (5.1.4) in the last step).

We have, therefore,

$$w_\alpha = e^{-(S_0 - S_{bath})/k} \qquad (5.1.10)$$

The entropy of the bath really depends only on its energy, which is $U_{bath} = U_0 - \varepsilon_\alpha$:

$$S^\alpha_{bath} = S_{bath}(U_0 - \varepsilon_\alpha) \qquad (5.1.11)$$

Because ε_α is much smaller than U_0, we can expand S_{bath} in a Taylor series:

$$S_{bath} = S_{bath}(U_0) - (\partial S_{bath}/\partial U_{bath})\varepsilon_\alpha \qquad (5.1.12)$$

But $\partial S_{bath}/\partial U_{bath}$ is just one over the temperature of the bath, T, so,

$$S^\alpha_{bath} = S_{bath}(U_0) - \varepsilon_\alpha/T \qquad (5.1.13)$$

Here $S_{bath}(U_0)$ is simply the entropy the bath would have if it had all of the system's energy U_0. It is not equal to S_0, and it does not have any other special significance.

By substituting Eq. (5.1.13) into Eq. (5.1.10) we get

$$w_\alpha = Ae^{-\varepsilon_\alpha/(k_B T)} \qquad (5.1.14)$$

where

$$A = e^{-(S_0 - S_{bath}(U_0))/(k_B T)} \qquad (5.1.15)$$

A depends on T, but it doesn't depend on the state α at all. In fact, since we know that $\sum_\alpha w_\alpha = 1$, applying this condition to Eq. (5.1.14) gives

$$A = 1 \Big/ \sum_\alpha e^{-\varepsilon_\alpha/(k_B T)}$$

In fact $A = 1/Z$, where

$$Z = \sum_\alpha e^{-\varepsilon_\alpha/(k_B T)}$$

is the **partition function**.

Problem 5.1

Is $S_{bath}(U_0)$ larger or smaller than S_0? Is it larger or smaller than the entropy of the bath when the system is in equilibrium? Explain.

We have now solved the problem that we posed at the outset. The probability of finding the sample in the state α with energy ε_α is just

$$w_\alpha = Ae^{-\varepsilon_\alpha/(k_B T)} \qquad (5.1.16)$$

where A depends on the temperature of the bath but is the same for all microscopic states of the sample, and $e^{-\varepsilon_a/(k_BT)}$ is called the Boltzmann factor.

To derive this very important result, we made remarkably few assumptions. We assumed the sample is very small compared with the bath, and we assumed the entropy of the bath depends only on its energy. These assumptions are surely true if the sample is an atom or a molecule, as we started out imagining, but it may also be true for a macroscopic sample. If the sample is, say, a box of ideal gas or a chunk of metal, the only conditions are that it be attached to the bath in such a way that N and V remain constant (otherwise the entropy of the bath will depend on N and V), that it be small compared with the bath (and we are free to imagine as large a bath as we wish), and that it has definite quantum states (all possible samples do). Thus the following statement is quite generally true:

> If any sample is at temperature T, the probability of finding it in the state α with energy ε_α is proportional to the Boltzmann factor, $e^{-\varepsilon_\alpha/(k_BT)}$.

Example 5.1.1

A paramagnetic salt is a crystalline solid in which there is a molecule that has a magnetic moment m at each site on the crystal lattice. According to the laws of quantum mechanics, if an external magnetic field $\vec{H}$ is applied, each magnetic moment in certain samples can have only two possible states: up (along the field) or down (opposite to the field). How does the magnetization of the salt depend on $|\vec{H}|$ and T at high temperature? What is meant by high temperature in this example?

Solution.

The energy of the molecule at each lattice site has a term, u_m, which depends on the orientation of the magnetic moment, given by

$$u_m = -\vec{m}\cdot\vec{H}$$

All other parts of the energy of the molecules are independent of the orientation of $\vec{m}$, and may be regarded as part of the temperature bath. Let us suppose there are N molecules per unit volume, of which N_1 have $\vec{m}$ pointing up with magnetic energy $-mH$ and N_2 have $\vec{m}$ pointing down with energy $+mH$. Then,

$$N_1/N = e^{-u_m(\text{up})/(k_BT)} \Big/ \sum_{\text{states}} e^{-u_m/(k_BT)}$$

$$= e^{+mH/(k_BT)} \Big/ \left(e^{mH/(k_BT)} + e^{-mH/(k_BT)} \right)$$

and

$$N_2/N = e^{-mH/(k_BT)} \bigg/ \left(e^{mH/(k_BT)} + e^{-mH/(k_BT)} \right)$$

Notice that $N_1 + N_2 = N$, as should be the case.

The magnetization (magnetic moment per unit volume) is

$$M = (N_1 - N_2)m$$
$$= Nm \tanh[mH/(k_BT)]$$

where

$$\tanh x = (e^x - e^{-x})/(e^x + e^{-x})$$

High temperature can only mean that k_BT is large compared with the only other energy in the problem, mH:

$$kT \gg mH$$

or

$$x = 1$$

In this limit, $e^x \simeq 1 - x$,

$$\tanh x = (1 + x) - (1 - x)/(1 + x) + (1 - x) = 2x/2 = x$$

So

$$M \simeq Nm[mH/(k_BT)] = Nm^2H/(k_BT)$$

This result is known as the Curie law for a paramagnetic salt.

Problem 5.2
The hydrogen atom consists of one electron bound to one proton. The electron has a lowest-energy state, called the ground state, and various higher-energy quantum states called excited states. The first (lowest-energy) excited state is 10.2 eV (electron volts) above the ground state. Find the temperature at which roughly 1% of hydrogen atoms will be in excited states.

Problem 5.3
There are various systems in nature (including the paramagnetic molecule in Example 5.1.1 above) that have two possible energy states. Let us call the two energies 0 and Δ (in the paramagnetic example above, if we defined the lower-energy state to be 0, the upper would be $2mH$). In terms of Δ, find the following.

(a) The quantity A as a function of T. Make a plot of A versus T.
(b) w_α for each of the two states. Sketch a plot of w_α versus Δ, regarding Δ as a continuous variable, for two different temperatures.
(c) The average energy as a function of T, $u(T)$.
(d) The rate of change of the average energy with temperature $C = du/dT$ (C is the specific heat). Show that $C(T)$ has a maximum and find the temperature of the maximum. Sketch a graph of $C(T)$ versus T.

5.2 What happened?

The result of the last section is that, for any sample having quantum states (and that means any sample whatever, so long as its N and V remain constant), the probability of finding it in any single quantum state α is

$$w_\alpha = Ae^{-\varepsilon_\alpha/(k_B T)} \qquad (5.2.1)$$

where ε_α is the energy of the sample in the state α, and A depends only on T. Equation (5.2.1) means that, at any given temperature, the lower the energy of the state, the more probable it is. This result raises two profound questions.

(1) We started with the assumption that all possible states are equally likely. How did we wind up with the result that lower-energy states are more probable than higher-energy states?
(2) According to Eq. (5.2.1), the ground state (i.e. the lowest-energy state) of any sample is always its most probable state. If that's true, why don't we commonly find macroscopic samples in their ground states? In fact, not only do we not commonly find macroscopic samples in their ground states, but it is actually one of the laws of thermodynamics (the third law), that it's impossible to reach the ground state of a macroscopic system. What's going on here?

Here are the answers:

(1) Our new result is based on the assumption that all possible quantum states of an isolated system are equally likely. To obtain the result, we first divided the system into two parts: the sample (very small compared with the rest) and the bath (the rest). For all possible quantum states of the system as a whole, the lower the energy of the sample, the greater the energy available to the bath. But because the bath is much bigger than the sample, it has many more ways to use up the extra energy than the sample would. In other words, the less energy in the sample and hence the more in the bath, the more

quantum states are available for the system as a whole, and therefore the more probable the situation is. That is what Eq. (5.2.1) means. The importance of the many states available in the bath entered our argument at Eq. (5.1.12), where we used the fact that $\partial S_{\text{bath}}/\partial U_{\text{bath}} = 1/T$. Temperature is the crucial point here. For any isolated system – here, meaning isolated from the rest of the Universe – having constant total energy, all quantum states are equally likely. However, if the system is a sample at constant temperature, which means it is able to exchange energy with a much larger bath, then instead it obeys Eq. (5.2.1). Under these conditions, lower-energy states are always more probable.

(2) A macroscopic sample has just one, single, unique ground state. That is one way of stating the third law of thermodynamics. At any higher energy, the sample may have many possible quantum states with the same energy. This is particularly true if the energy is a little fuzzy (uncertain). As we've seen, that's always the case in the real world. Then the sample will generally have many possible states at nearly the same energy. Each of these states is less probable than the ground state, but there are so many of them that the sample is much more likely to be found in one of those than it is to be found in the ground state. We will explore this crucial idea quantitatively in the next section.

5.3 The density of states

To compute the average or thermodynamic energy of a sample at temperature T (which we will understand technically to mean in contact with a temperature bath at temperature T) we use Eq. (5.1.7) together with Eq. (5.2.1), and write

$$\bar{\varepsilon} = A \sum_{\text{states}} \varepsilon_\alpha e^{-\varepsilon_\alpha/(k_{\text{B}}T)} \tag{5.3.1}$$

Performing this sum is a formidable task because there are many quantum states for any macroscopic sample. We can make progress, however, by making use of the insight at the end of the last section: The number of states of roughly the same energy will be an important consideration. This suggests we rearrange the sum in Eq. (5.3.1):

$$\sum_{\text{states}} \rightarrow \sum_{\text{energies}} (\text{number of states per small range of energy})$$

$$\times (\text{range of energies}) \tag{5.3.2}$$

Note that Eq. (5.3.2) is not really an equation. $\sum_{\text{states}}$ is not equal to anything; it is an operation that will be replaced by a new operation, and, before it becomes an equation, it has to operate on something. The new operation will involve collecting states into bins of roughly the same energy.

To make the operation more quantitative, we'll suppose the sample has energy ε, with uncertainty $\Delta\varepsilon$ around ε. Then,

$$\text{(number of states in } \Delta\varepsilon) = \rho(\varepsilon)\Delta\varepsilon \tag{5.3.3}$$

Here $\rho(\varepsilon)$ is called the *density of states*. In Eq. (5.3.2) $\rho(\varepsilon)$ is the first term in parentheses and $\Delta\varepsilon$ is the second. With this definition, Eq. (5.3.1) becomes

$$\bar{\varepsilon} = A\sum_{\Delta\varepsilon} \varepsilon\rho(\varepsilon)e^{-\varepsilon/(k_B T)}\,\Delta\varepsilon \tag{5.3.4}$$

where we are summing over energy bins of width $\Delta\varepsilon$. It will nearly always turn out to be easier to perform the sum as an integral, shrinking $\Delta\varepsilon$ to an infinitesimal, $d\varepsilon$,

$$\bar{\varepsilon} = A\int \varepsilon\rho(\varepsilon)e^{-\varepsilon/(k_B T)}\,d\varepsilon \tag{5.3.5}$$

Performing the same operations on Eq. (5.2.1) gives

$$A = 1 \Big/ \int_0^\infty \rho(\varepsilon)e^{-\varepsilon/(k_B T)}\,d\varepsilon \tag{5.3.6}$$

Thus we have very nearly reduced the entire problem of statistical physics to finding $\rho(\varepsilon)$ for any given sample.

Let us find $\rho(\varepsilon)$ for a single atom in an ideal gas. The energy levels that may be occupied by an atom of ideal gas were given in Eqs. (1.2.1)–(1.2.4). We repeat them here:

$$\varepsilon = p^2/(2m) \tag{1.2.1}$$

where

$$p^2 = p_x^2 + p_y^2 + p_z^2 \tag{1.2.2}$$

and

$$\begin{aligned} p_x &= n_x p_0 \\ p_y &= n_y p_0 \\ p_z &= n_z p_0 \end{aligned} \tag{1.2.3}$$

where n_x, n_y, $n_z = 0, \pm 1, \pm 2, \ldots$, and

$$p_0 = h/L \qquad (1.2.4)$$

In these equations, ε is energy, p is momentum, m is the mass of the atom and $L = V^{1/3}$ is the size of the box the atoms are in.

The trick to finding $\rho(\varepsilon)$ is to imagine a new kind of space called momentum space or p-space. The coordinates in this space are p_x, p_y and p_z. In classical physics (where $\vec{p}$ is a continuously variable vector), any point in p-space, (p_x, p_y, p_z), represents a possible state of an atom whose total momentum would be given by Eq. (1.2.2), and whose energy would be given by Eq. (1.2.1).

As the atom bounces around in the box in real space, exchanging energy and momentum with the walls and the other atoms, it is also bouncing around from point to point in p-space.

> **Problem 5.4**
> A classical atom in state (p_x, p_y, p_z) undergoes an elastic collision with a wall parallel to the y, z plane. What are its new coordinates in p-space? Make a sketch of p-space showing the two points (before and after).

In quantum mechanics, unlike classical mechanics, (p_x, p_y, p_z) has only discrete allowed values, which are given by Eqs. (1.2.3) and (1.2.4). This is the property that makes the states countable. We can visualize where in p-space the allowed states are found by laying out a grid whose spacing is p_0. Each corner of this three-dimensional grid is a point that can be reached from the origin by a vector

$$\vec{p} = (p_x, p_y, p_z) = p_0(n_x, n_y, n_z)$$

where each n is zero or a positive or negative integer.

We now ask the following question: How many states are there in the energy range from zero to some definite value ε? The question would make no sense at all in classical physics, but in quantum mechanics, the answer is easy. At energy ε, an atom has momentum $p = |\vec{p}| = \sqrt{2m\varepsilon}$. All states with energy less than ε are found inside a sphere in p-space whose equation is, as we noted above, Eq. (1.2.2),

$$p^2 = p_x^2 + p_y^2 + p_z^2$$

(This equation is perfectly analogous to the equation of a sphere of radius R in real space, $R^2 = x^2 + y^2 + z^2$.)

The number of points inside this sphere is just equal to the number of grid points inside. Provided that the sphere is very large compared with the spacing of the grid, we can write

(number of states inside the sphere) = (volume of the sphere)/
$$\qquad\qquad\qquad\qquad\qquad\text{(volume of one cube)}$$

$$= \frac{4}{3}\pi p^3/p_0^3 \qquad\qquad (5.3.7)$$

The density of states, $\rho(\varepsilon)$, is the number of states in a spherical shell of radius $p = \sqrt{2m\varepsilon}$ and thickness dp,

(number of states in spherical shell) $= \rho(\varepsilon)d\varepsilon = 4\pi p^2 \, dp/p_0^3$ (5.3.8)

It's easy to see that this gives us back Eq. (5.3.7):

$$\text{(number of states with energy from zero to } \varepsilon\text{)} = \int_0^\varepsilon \rho(\varepsilon')d\varepsilon'$$

$$= \int 4\pi p'^2 \, dp'/p_0^3$$

$$= \frac{4}{3}\pi p^3/p_0^3$$

where p is related to ε by $\varepsilon = p^2/(2m)$ (in the integral above we have used ε' and p' to denote dummy variables). Going back to Eq. (5.3.8), we have

$$\rho(\varepsilon) = (4\pi p^2)(dp/d\varepsilon)/p_0^3$$

Substituting in $p^2 = 2m\varepsilon$, $dp/d\varepsilon = \sqrt{m/(2\varepsilon)}$ and $p_0^{-3} = V/h^3$, we have

$$\rho(\varepsilon) = \frac{4\pi\sqrt{2\varepsilon}}{h^3}Vm^{3/2} \qquad\qquad (5.3.9)$$

This result is the density of states of a single atom in an ideal gas.

Problem 5.5
Find a formula for the number of states of a single atom in a box 1 cm on each side, with energies between zero and k_BT, where T is room temperature.

Problem 5.6
Find the density of states for an atom in a two-dimensional box.

Problem 5.7
Suppose the particles in a box obey Eqs. (1.2.2) and (1.2.3), but instead of Eq. (1.2.1), energy and momentum are related by $\varepsilon = c|\vec{p}|$ (this is the case for photons). Find $\rho(\varepsilon)$.

5.4 Energy and entropy

Now that we have $\rho(\varepsilon)$ for an atom in an ideal gas, we can use it to calculate the average energy, Eq. (5.3.5),

$$\bar{\varepsilon} = A \int_0^\infty \varepsilon \rho(\varepsilon) e^{-\varepsilon/(k_B T)} \, d\varepsilon$$

where (Eq. (5.3.6))

$$A = 1 \bigg/ \int_0^\infty \rho(\varepsilon) e^{-\varepsilon/(k_B T)} \, d\varepsilon$$

Here, $\rho(\varepsilon) \propto \varepsilon^{1/2}$ and $\varepsilon\rho(\varepsilon) \propto \varepsilon^{3/2}$. All the other terms cancel out when we insert the expression for A, leaving

$$\bar{\varepsilon} = \frac{\displaystyle\int_0^\infty \varepsilon^{3/2} e^{-\varepsilon/(k_B T)} \, d\varepsilon}{\displaystyle\int_0^\infty \varepsilon^{1/2} e^{-\varepsilon/(k_B T)} \, d\varepsilon} \qquad (5.4.1)$$

Here it's useful to change variables, writing $\varepsilon/(k_B T) = x$. This yields

$$\bar{\varepsilon} = kT_B \frac{\displaystyle\int_0^\infty x^{3/2} e^{-x} \, dx}{\displaystyle\int_0^\infty x^{1/2} e^{-x} \, dx} \qquad (5.4.2)$$

The integrals are what are known as gamma functions (no relation to our Γ, the number of possible states of a system),

$$\Gamma(n+1) = \int x^n e^{-x} \, dx \qquad (5.4.3)$$

Gamma functions are generalized factorials, having the property

$$\Gamma(n + 1) = n\Gamma(n) \qquad (5.4.4)$$

Our expression for $\bar{\varepsilon}$ becomes

$$\bar{\varepsilon} = kT_B \frac{\Gamma(5/2)}{\Gamma(3/2)} = \frac{3}{2} kT_B \qquad (5.4.5)$$

If an ideal gas has N atoms with average energy $\bar{\varepsilon} = (3/2)kT_B$, the total energy is

$$U = \frac{3}{2} Nk_B T \qquad (5.4.6)$$

Equations (5.4.5) and (5.4.6) are examples of a rule known as the **equipartition of energy**. A gas atom in three dimensions is said to have three degrees of freedom (the number of components of

$p^2 = p_x^2 + p_y^2 + p_z^2$). N atoms have between them $3N$ degrees of freedom. According to the equipartition of energy, any thermodynamic system has an average energy of $(1/2) k_B T$ for each degree of freedom, where a degree of freedom means a quadratic term in the energy (p_x^2, p_y^2 and p_z^2 for each atom). If the system has potential energy that depends on components x^2, y^2 and z^2, those too count as quadratic degrees of freedom. The ideal gas, however, does not have potential energy (the energy does not depend on the positions of the atoms so long as they are somewhere in the box), so it has $3N$ degrees of freedom, and its energy is given by Eq. (5.4.6).

We have derived the equipartion of energy of an ideal gas starting from its quantum states. However, equipartition is really a classical result that is not always true in quantum systems. It is true only if $k_B T$ is very large compared with the spacing between energy states. The same approximation allowed us to turn the sum over states, Eq. (5.3.4), into an integral over states, Eq. (5.3.5). We could not have let the interval between states (the uncertainty in the energy of a state), $\Delta\varepsilon$, shrink to an infinitesimal, $d\varepsilon$, if it were not true that $k_B T \gg \Delta\varepsilon$.

Problem 5.8
Find the energy of a two-dimensional ideal gas (assuming equipartition of energy).

Problem 5.9
To a good approximation, a crystalline solid may be regarded as a collection of N three-dimensional harmonic oscillators, where, for each dimension (say, for motion parallel to the x axis), the energy is given by

$$\varepsilon = \frac{p_x^2}{2m} + \frac{1}{2}kx^2$$

where k is the spring constant. In this approximation, find the average energy, U, of a solid at high temperature.

Problem 5.10
Sketch graphs of $\rho(\varepsilon)$ and $\varepsilon\rho(\varepsilon)$ versus $\varepsilon/(k_B T)$ for an atom in an ideal gas. Use these to explain (quantitatively) why, although the most probable state is $\varepsilon = 0$, the average energy is $\bar\varepsilon = (3/2)k_B T$.

We can find the entropy of any sample by means of an ingenious trick. The entropy of the system is given by

$$S_0 = S_{\text{bath}} + S \tag{5.4.7}$$

where S_{bath} and S are the equilibrium entropies of the bath and the sample, respectively. For each quantum state, α, of the sample, the bath has entropy S_{bath}^α given by Eq. (5.1.9),

$$S_0 - S_{bath}^\alpha = -k_B \log w_\alpha \qquad (5.4.8)$$

Like for any other quantity having a definite value for each state of the sample, we can find the thermodynamic average value by averaging over the states of the sample using Eq. (5.1.6), with $f = S_0 - S_{bath}^\alpha$:

$$\begin{aligned} \overline{S}_0 - \overline{S}_{bath}^\alpha &= \sum_\alpha (S_0 - S_{bath}^\alpha) w_\alpha \\ &= -k_B \sum_\alpha w_\alpha \log w_\alpha \end{aligned} \qquad (5.4.9)$$

But the average value of S_{bath}^α is the thermodynamic equilibrium entropy of the bath, S_{bath} in Eq. (5.4.7). Thus $\overline{S}_0 - \overline{S}_{bath}^\alpha = S_0 - S_{bath} = S$ according to Eq. (5.4.7). Thus,

$$S = -k_B \sum_\alpha w_\alpha \log w_\alpha \qquad (5.4.10)$$

This important result is true for any sample.

Problem 5.11
Find the entropy at temperature T of a two-state system such as that in Problem 5.3 at all temperatures. What is the limiting value of the entropy at very high temperature?

Let us use Eq. (5.4.10) to find S_1, the entropy of a single atom of an ideal gas:

$$\begin{aligned} S_1 &= -k_B \sum_\alpha w_\alpha \log w_\alpha \\ &= -k_B A \int_0^\infty e^{-\varepsilon/(k_B T)} \log\left(A e^{-\varepsilon/(k_B T)} \right) \rho(\varepsilon) d\varepsilon \end{aligned} \qquad (5.4.11)$$

where we have in one step used Eq. (5.1.14) for w_α and converted the sum to an integral in the usual way:

$$\sum_\alpha (\text{something with subscript } \alpha) = \int (\text{continuous variable}) \rho(\varepsilon) d\varepsilon \quad (5.4.12)$$

In the integrand in Eq. (5.4.11),

$$\log(A e^{-\varepsilon/(k_B T)}) = \log A - \varepsilon/(k_B T) \qquad (5.4.13)$$

and log A doesn't depend on ε. Thus,

$$S_1 = -k_B A \log A \int_0^\infty e^{-\varepsilon/(k_B T)} \rho(\varepsilon) d\varepsilon$$

$$= -k_B A \int_0^\infty e^{-\varepsilon/(k_B T)} [-\varepsilon/(k_B T)] \rho(\varepsilon) d\varepsilon \qquad (5.4.14)$$

The first term simplifies because, according to Eq. (5.1.16),

$$A \int e^{-\varepsilon/(k_B T)} \rho(\varepsilon) d\varepsilon = 1$$

and the second term is, according to Eq. (5.3.5),

$$\frac{kA}{kT} \int_0^\infty \varepsilon e^{-\varepsilon/(k_B T)} \rho(\varepsilon) d\varepsilon = \frac{\bar{\varepsilon}}{k_B T}$$

where $\bar{\varepsilon}(3/2)k_B T$. Thus

$$S_1 = -k_B \log A + 3k_B/2 \qquad (5.4.15)$$

Problem 5.12
Show that, for any sample, $k_B T \log A = U - TS$.

The quantity $U - TS$ is the **Helmholtz free energy**,

$$F = U - TS \qquad (5.4.16)$$

and Z, the partition function, is given by

$$Z = 1/A \qquad (5.4.17)$$

or

$$\log Z = -\log A \qquad (5.4.18)$$

The resulting equation,

$$F = -k_B T \log Z \qquad (5.4.19)$$

is one of the most important and useful results in statistical mechanics.
For a single atom in an ideal gas, we still have to find the quantity

$$Z_1 = 1/A = \int_0^\infty \rho(\varepsilon) e^{-\varepsilon/(k_B T)} d\varepsilon$$

$$= a \int_0^\infty \varepsilon^{1/2} e^{-\varepsilon/(k_B T)} d\varepsilon \qquad (5.4.20)$$

$$= (k_B T)^{3/2} a \int_0^\infty x^{1/2} e^{-x} dx$$

where $a = 4\pi\sqrt{2}Vm^{3/2}/h^3$. In the first step, we have used Eq. (5.3.9) and in the second we have changed variables, $x = \varepsilon/(k_BT)$. With the help of a table of definite integrals, we find

$$\int x^{1/2}e^{-x}\,dx = \sqrt{\pi}/2 \qquad (5.4.21)$$

So

$$Z_1 = (k_BT)^{3/2}\frac{2\sqrt{2\pi^3}Vm^{3/2}}{h^3} \qquad (5.4.22)$$

Z_1 is the partition function of a single atom in an ideal gas. Substituting into Eq. (5.4.15) with $\log Z_1 = -A$ gives

$$S_1 = \frac{3}{2}k_B\log\left[m\pi k_BT\left(\frac{2\sqrt{2V^2}}{h^3}\right)^{3/2}\right] + \frac{3}{2}k_B \qquad (5.4.23)$$

Problem 5.13
Show that Eq. (5.4.23) cannot be correct in the limit $T \to 0$. Explain why, and find an expression for the entropy in that limit.

5.5 The entropy of an ideal gas

It is tempting to suppose that, just as the energy of an ideal gas is $U = N\bar{\varepsilon}$, the entropy should be NS_1. That supposition is wrong, however, for the reason discussed at the end of Section 1.2: If Γ_1 is the number of possible states of one atom, then,

$$S_1 = k_B\log\Gamma_1 \qquad (5.5.1)$$

and

$$NS_1 = Nk_B\log\Gamma_1 = k_B\log\Gamma_1^N$$

but the number of possible states of N atoms of an ideal gas is much smaller than Γ_1^N because the atoms are indistinguishable. We would have Γ_1^N states if we could specify each state by saying atom a is in level α_1, atom b is in level α_2 and so on. As we saw in Section 1.2, that is the wrong way of counting states. We cannot say *which* atoms are in each level; we can only say *how many* are in each level.

The problem is surprisingly easy to deal with, provided the temperature is high enough or the density low enough that there is little chance that more than one atom will be found in a single level (a level, remember is the set of three positive or negative integers, n_x, n_y and n_z, that specify a quantum state of one atom). Under these circumstances, instead of saying that atom a is in level α_1, etc., we can say "There is an atom in level α_1. It can be any of the N atoms in the box. There is an atom in level α_2. It could be any of the remaining $N - 1$ atoms." And so on, for every possible level. Thus the original, mistaken method of counting states gave too many by $N \times (N - 1) \times (N - 2) \times \cdots = N!$ That's because we counted too many states by a factor N when we said each atom instead of any atom could be in state α_1, each atom could be in state α_2 and so on. Thus the correct value of Γ is simply given by

$$\Gamma = \frac{\Gamma_1^N}{N!} \qquad (5.5.2)$$

So the real entropy of an ideal gas is

$$S = k_B \log \Gamma$$
$$= N k_B \log \Gamma_1 - k_B \log N!$$
$$= N S_1 - k_B \log N! \qquad (5.5.3)$$

To evaluate this result we use what is known as ***Stirling's approximation***,

$$\log N! = N \log N - N \qquad (5.5.4)$$

Problem 5.14
Test Sterling's approximation for a few small numbers (until you get tired of calculating factorials).

Stirling's approximation is extremely accurate for the large N of a real ideal gas ($N \sim 10^{23}$, say). Using it, we find

$$S = N S_1 - N k_B \log N + N k_B$$

$$= \frac{5}{2} N k_B + \frac{3}{2} N k_B \log(\pi m k_B T) + N k_B \log\left(\frac{2V\sqrt{2}}{h^3}\right) - N k_B \log N$$

$$= N k_B \log\left[\frac{(\pi m k_B T)^{3/2}}{N h^3} e^{5/2} 2\sqrt{2V^2}\right] \qquad (5.5.5)$$

With just a bit of manipulation we find that

$$\Gamma = \left[(2\pi m k_B T / h^2)^{3/2} \frac{e^{5/2}}{N/V} \right]^N \qquad (5.5.6)$$

This result is the same as Eq. (1.5.2), and justifies Eq. (1.1.4), which was the starting point of the ideal gas example given in this book.

Problem 5.15

Starting from Eqs. (5.4.6) and (5.5.5), derive Eq. (1.1.4).

6

Thermodynamic functions

6.1 Free energies

Let us begin by recalling the **Helmholtz free energy**,

$$F = U - TS \qquad (6.1.1)$$

Then,

$$dF = dU - T\,dS - S\,dT$$

or

$$dF = -S\,dT - P\,dV \qquad (6.1.2)$$

So, F is a function of T and V, $F = F(T, V)$ or $F = F(T, V, N)$. T and V are said to be the proper variables of F. If F is given as a function of T, V and something else, its equilibrium value will be found by minimizing it with respect to the something else. We have, from Eq. (6.1.2),

$$S = -(\partial F/\partial T)_V, \qquad P = -(\partial F/\partial V)_T \qquad (6.1.3)$$

Taking the second derivative, we have

$$\frac{\partial^2 F}{\partial T\,\partial V} = \left(\frac{\partial S}{\partial V}\right)_T = \left(\frac{\partial P}{\partial T}\right)_V$$

so

$$(\partial S/\partial V)_T = (\partial P/\partial T)_V \qquad (6.1.4)$$

Equation (6.1.4) is called a **Maxwell relation**.

In a similar manner to the free energy we can define the **enthalpy**,

$$H = U + PV$$

Then,

$$dH = dU + P\,dV + V\,dP$$

or

$$dH = T\,dS + V\,dP \tag{6.1.5}$$

so H is a function of entropy and pressure. For example it might measure how heat spreads in a flow process that takes place at constant pressure. We have from Eq. (6.1.5),

$$T = (\partial H/\partial S)_P, \qquad V = (\partial H/\partial P)_S$$

with second derivatives

$$\frac{\partial^2 H}{\partial S\,\partial P} = \left(\frac{\partial V}{\partial S}\right)_P = \left(\frac{\partial T}{\partial P}\right)_S$$

so we have another Maxwell relation,

$$(\partial V/\partial S)_P = (\partial T/\partial P)_S \tag{6.1.6}$$

Finally, we can define one more function, called the **Gibbs function** or sometimes the **Gibbs free energy**,

$$G = U - TS + PV \tag{6.1.7}$$

so that $dG = dF + p\,dV + V\,dP$, or

$$dG = -S\,dT + V\,dP \tag{6.1.8}$$

Thus the proper variables of G are temperature and pressure, and we have $S = -(\partial G/\partial T)_P$ and $V = (\partial G/\partial P)_T$. On taking the second derivatives we have another Maxwell relation,

$$(\partial V/\partial T)_P = -(\partial S/\partial P)_T \tag{6.1.9}$$

The question arises, how can we remember this blizzard of equations?

The key to the answer is to make a box with letters and arrows as in Fig. 6.1.

There is a mnemonic for remembering the letters, starting at the midpoint of the right-hand side and proceeding clockwise around the perimeter: Good Physicists Have Studied Under Very Fine Teachers.

(1) Each energy function at the midpoint of each side is flanked by its proper variables: $F = F(T,V)$, $G = G(T,P)$, $H = H(P,S)$ and $U = U(S,V)$.
(2) For differentials, take each energy function, go to the opposite corners and across the diagonals, following the sign convention

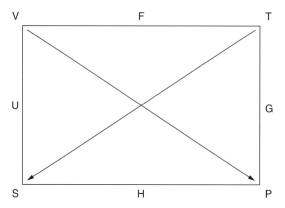

Figure 6.1 The mnemonic of the energy functions.

plus, if your path goes toward the arrow, or *minus*, if it does not:
$dF = -S\,dT - P\,dV$, $dG = -S\,dT + V\,dP$ and so on.

(3) Transformations: For each energy function, go to an adjoining energy
function, then across the next diagonal, following the sign convention: $G = H - TS$; or, going the other way, $G = F + PV$, and so on.

(4) Maxwell relations: Starting from one corner, go around three corners
of the box in either direction and then around the opposite three
corners, ending up on the same side as your starting point and
obeying the sign on the final side: $(\partial V/\partial T)_P = -(\partial S/\partial P)_T$, and so on.

Example 6.1.1
Use the mnemonic box to produce all four of the above results.

---

(1) We already have these.
(2) We already have dF and dG. The others are $dH = T\,dS + V\,dP$ and
$dU = T\,dS + P\,dV$.
(3) We already have G; the others are $H = U + PV$, or $H = G + TS$;
$U = F + TS$, or $U = H - PV$; and $F = G - PV$, or $F = U - TS$.
(4) We already have one; the other three are $(\partial T/\partial P)_S = (\partial V/\partial S)_P$,
$(\partial P/\partial S)_V = (\partial T/\partial V)_S$, and $(\partial P/\partial T)_V = (\partial S/\partial V)_T$.

If you spend 15 minutes practicing these steps, they may be yours forever.

Example 6.1.2
The Joule–Thomson effect.

See Problem 2.19. A piston pushes a gas at pressure P_1 through a nozzle
into a second chamber at lower pressure, P_2, where a second piston on

the other side of the nozzle goes out to keep the pressure constant. We proved earlier that $U_1 + P_1V_1 = U_2 + V_2P_2$, i.e. that $\Delta H = 0$. In this case, find an expression for the resulting change in temperature, i.e. for $J = (\partial T/\partial P)_H$.

Setting $dH = T\,dS + V\,dP$ with $dH = 0$ and "dividing" by dT gives

$$-T(\partial S/\partial T)_H = (V\,\partial P/dT)_H$$

or

$$(\partial T/\partial P)_H = -\frac{V}{T(\partial S/\partial T)_H}$$

Now write, for $S(T,P)$, $dS = (\partial S/\partial T)_P\,dT + (\partial S/\partial P)_T\,dP$, so that

$$(\partial S/dT)_H = (\partial S/\partial T)_P + (\partial S/\partial P)_T + (\partial P/\partial T)_H$$

We have $J = -(V/T)(\partial S/\partial T)_H$ and

$$\left(\frac{\partial S}{\partial T}\right)_H = \frac{C_P}{T} + \frac{1}{J}\left(\frac{\partial S}{\partial P}\right)_T$$

so $-V = JT(\partial S/\partial T)_H = JC_P + T(\partial S/\partial P)_T$, or

$$J = -\frac{V + T(\partial S/\partial P)_T}{C_P}$$

The problem here is that $(\partial S/\partial P)_T$ is not very useful. We look at our thermodynamic mnemonic and find

$$(\partial S/\partial P)_T = -(\partial V/\partial T)_P$$

and thus, finally,

$$J = \frac{V}{C_P}(\beta T - 1)$$

where

$$\beta = \frac{1}{V}(\partial V/\partial T)_P$$

Maxwell's relations simply express the fact that $U(S,V)$ exists and is a well-behaved function for a given N. It then follows that $F(T,V) = U - TS$ also exists, and so on. In other words, if $f(x,y)$ exists and is a well-behaved function, then

$$\frac{\partial^2 f}{\partial x\,\partial y} = \frac{\partial^2 f}{\partial y\,\partial x} \tag{6.1.10}$$

For example, suppose

$$f(x,y) = 3x^2 + y^2x$$

Then, $\partial f/\partial x = 6x + y^2$, so

$$\frac{\partial^2 f}{\partial y\,\partial x} = 2y$$

and $\partial f/\partial y = 2yx$, so

$$\frac{\partial^2 f}{\partial x\,\partial y} = 2y$$

In general, if

$$df = M\,dx + N\,dy \qquad (6.1.11)$$

then $M = (\partial f/\partial x)_y$ and $N = (\partial f/\partial y)_x$, and we must have

$$(\partial M/\partial y)_x = (\partial N/\partial x)_y \qquad (6.1.12)$$

We would like to say that, if df is integrable, then $f(x,y)$ exists and is well behaved. That isn't always true, but it is always true for the thermodynamic variables of a system in equilibrium. The existence of a well-behaved function $U(S,V)$ was the starting point of our discussion.

Example 6.1.3

Find J for an ideal gas and for a gas obeying the leading-order virial correction,

$$PV = Nk_BT\left[1 + \frac{N}{V}B(T)\right]$$

We have

$$J = \frac{V}{C_P}(\beta T - 1)$$

For the ideal gas, $\beta = 1/T$, so $J = 0$. For the virial gas,

$$V = \frac{Nk_BT}{P} + NB(T)$$

so

$$\left(\frac{\partial V}{\partial T}\right)_P = \frac{Nk_B}{P} + N\frac{dB}{dT}$$

and

$$J = \frac{1}{C_P}\left[T\left(\frac{\partial V}{\partial T}\right)_P - V\right]$$

So we get

$$J = \frac{N}{C_P}\left[T\frac{dB}{dT} - B(T)\right]$$

It turns out that dB/dT is always positive, but $B(T)$ is negative at low temperature and positive at high temperature. Thus J can be either positive or negative, depending on the temperature.

Problem 6.1
Find J for a gas obeying the van der Waals equation of state, Eq. (3.2.5).

6.2 The chemical potential

Up to now, we've worked with fixed N. Now let's let N vary as well. Recalling Eq. (1.1.3), we have

$$dU = (\partial U/\partial S)_{V,N}\, dS + (\partial U/\partial V)_{S,N}\, dV + (\partial U/\partial N)_{S,V}\, dN$$

and, given Eqs. (1.1.5), (1.1.6) and (1.1.7),

$$dU = T\, dS - P\, dV + \mu\, dN \qquad (6.2.1)$$

where (Eq. (1.1.7))

$$\mu = (\partial U/\partial N)_{S,V}$$

is called the *chemical potential*. We still have $F = U - TS$, so

$$dF = -S\, dT - P\, dV + \mu\, dN \qquad (6.2.2)$$

so that $F = F(T,V,N)$ and

$$\mu = (\partial F/\partial N)_{T,V} \qquad (6.2.3)$$

Likewise, $H = U + PV$, so $dH = T\, dS + V\, dP + \mu\, dN$,

$$\mu = (\partial H/\partial N)_{S,P} \qquad (6.2.4)$$

and, furthermore, $G = F + PV$, so that $dG = -S\, dT + V\, dP + \mu\, dN$, and

$$\mu = (\partial G/\partial N)_{T,P} \qquad (6.2.5)$$

Thus μ can be found by taking a partial derivative with respect to N of any one of the energy functions.

As we saw earlier, S, V and N (as well as all the energy functions) are said to be **extensive variables**; that is, they scale with the size of the system. On the other hand, T, V and μ are **intensive variables**; that is, they do not depend on the size of the system.

We understand the meaning of T and V, but we know close to nothing about μ. We can do a little investigating to find out how μ behaves. Consider a box, with a certain U, because it has definite values of S, V and N. Now suppose there are λ such boxes, having λU, λS, λV and λN. Then,

$$d(\lambda U) = T\, d(\lambda S) + P\, d(\lambda V) + \mu\, d(\lambda N)$$

or

$$\lambda\, dU + U\, d\lambda = \lambda(T\, dS + P\, dV + \mu\, dN) + (TS + PV + \mu N)d\lambda$$

But λ is completely arbitrary. This must be true for any λ. Thus,

$$dU = T\, dS - P\, dV + \mu\, dN$$

which we already know to be true, and

$$U = TS - PV + \mu N \qquad (6.2.6)$$

which is a new result. And, since $G = F + PV = U - TS + PV$, we have, from Eq. (6.1.18),

$$G = \mu N \qquad (6.2.7)$$

So

$$dG = \mu\, dN + N\, d\mu$$

But we also had $dG = -S\, dT + V\, dp + \mu\, dN$, so it follows that

$$d\mu = -\frac{S}{N} dT + \frac{V}{N} dP \qquad (6.2.8)$$

When we add particles to any system, each of that system's energy functions changes by $\mu\, dN$. Does that mean that μ must be positive? The answer is no! Consider an ideal gas. We have $\mu = (\partial U / \partial N)_{S,\, V}$. If we add an atom, U must decrease in order to keep S constant. Thus, μ is negative. In general, it can be either positive or negative.

Consider a box made up of atoms or molecules, with fixed N at constant T and V. Let's divide the box up with an imaginary partition into sides 1 and 2 with N_1 and V_1 in side 1 and N_2 and V_2 in side 2, so that $N_1 + N_2 = N$, and $V_1 + V_2 = V$. N_1 will be determined by

$$\partial F/\partial N_1 = 0 = \partial(F_1 + F_2)/\partial N_1 = \partial F_1/\partial N_1 - \partial F_2/\partial N_2$$

or

$$\mu_1 - \mu_2 = 0 \tag{6.2.9}$$

But the position of the partition is arbitrary, so it follows that μ is uniform everywhere in the box.

Now imagine that, instead of a uniform phase in the box, we have an interface between a liquid, with chemical potential μ_l, and a gas, with chemical potential μ_g. Equilibrium between liquid and gas takes place at a given pressure for each temperature, so we have

$$\mu_l(P,T) = \mu_g(P,T) \tag{6.2.10}$$

This will be the equation of the liquid–gas coexistence curve in the $P - T$ plane. In particular, $(\partial \mu_l/\partial P)_{coex} \, dP + (\partial \mu_l/\partial T)_{coex} \, dT$ is equal to the same expression for μ_g. On putting this into Eq. (6.2.8), we have

$$(s_l - s_g)dT = (v_g - v_l)dP$$

where $S/N = s$ and $V/N = v$. Thus, finally,

$$(dP/dT)_{coex} = \frac{s_l - s_g}{v_g - v_l} \tag{6.2.11}$$

Equation (6.2.11) is known as the **Clausius–Clapeyron equation**.

The latent heat of evaporation is given by

$$l = T(s_g - s_l) \tag{6.2.12}$$

Thus,

$$(dP/dT)_{coex} = l/[T(v_g - v_l)] \tag{6.2.13}$$

Usually, $v_g \gg v_l$, and, since

$$v_g = \frac{V_g}{N_g} = \frac{k_B T}{P}$$

we have

$$l = \frac{k_B T^2}{P} \frac{dP}{dT} = -k \frac{d\log P}{d(1/T)} \tag{6.2.14}$$

A similar argument applied along the melting curve, where μ_s is the chemical potential of the solid, gives

$$\mu_s(P,T) = \mu_l(P,T) \qquad (6.2.15)$$

and

$$(dP/dT)_{\text{melting}} = \frac{s_l - s_g}{v_l - v_g} \qquad (6.2.16)$$

It is possible to have three-phase equilibrium, given by the equations

$$\mu_l(P,T) = \mu_g(P,T) = \mu_s(P,T) \qquad (6.2.17)$$

Since these are two equations in two unknowns, the solution exists at a single point, which we identified earlier as the triple point. The curve dP/dT at gas–liquid coexistence always has a positive slope, but at melting, v_l can be greater or less than v_s, so $(dP/dT)_{\text{melting}}$ can slope either way.

We have one more energy function to define, called the **Landau potential**,

$$\Omega = F - \mu N \qquad (6.2.18)$$

Then we have

$$d\Omega = dF - \mu \, dN - N \, d\mu = -S \, dT - P \, dV - N \, d\mu \qquad (6.2.19)$$

So $\Omega = \Omega(T,V,\mu)$ and

$$N = -(\partial\Omega/\partial\mu)_{T,V} \qquad (6.2.20)$$

Notice that

$$\Omega = F - \mu N = U - TS - \mu N = -PV$$

So, just as $G = \mu N$ (Eq. (6.2.7)),

$$\Omega = -PV \qquad (6.2.21)$$

6.3 Variational principles

If we have a box of matter with a certain U, V and N, it will have a number of possible states, Γ, which, if the system is not yet in equilibrium, may be a function of time. Equilibrium occurs when Γ reaches its maximum possible value, or, given Eq. (1.1.1), when $S = S_{\text{max}}$, the entropy, reaches its maximum possible value. In other words, as an isolated macroscopic system, not yet in equilibrium, goes from state A to state B, we should always have $S(B) \geq S(A)$. S will continue to increase until it can't get any bigger, at which point equilibrium has been reached. That means, for a system not yet in equilibrium, assuming we can nevertheless assign a temperature to it, $dS \geq dU/T$, with the equals sign obtaining at equilibrium. Generalizing to systems in which N and V can also fluctuate, we have

$$dU \leq T\,dS - P\,dV + \mu\,dN \qquad (6.3.1)$$

Here's how to use this equation. Suppose we have a box of stuff in a temperature bath (T constant). Then $d(U - TS) \leq 0$, or, in other words, $U - TS = F$ will shrink until equilibrium is achieved. To put it more succinctly,

$$\delta(U - TS) \leq 0$$

Or

$$\delta F \leq 0$$

So, F must be a minimum in equilibrium.

> **Problem 6.2**
> Suppose we have a box of gas in a temperature bath (T, V and N constant). Let the box have an imaginary partition dividing it into V_1 and V_2 such that $V_1 + V_2 = V =$ constant. What do we learn by minimizing the free energy with respect to V_1? (Remember that, when we minimize a function, the first derivative is equal to zero and the second derivative is positive.)

Suppose that, instead of T, V, $N =$ constant, we have T, P, $N =$ constant. Then Eq. (6.3.1) becomes

$$\delta(U - TS + PV) \leq 0$$

or, in other words,

$$\delta G \leq 0$$

So, in these conditions G must be a minimum.

> **Problem 6.3**
> Find the conditions for U and H to be minima.

6.4 Phase equilibrium in systems of more than one constituent

Suppose we have a mixture of more than one constituent. Each constituent will have its own chemical potential, since the chemical potential always refers to a particular, distinguishable substance. Thus we have

$$dU = T\,dS - P\,dV + \mu_1\,dN_1 + \mu_2\,dN_2 + \cdots \text{(etc.)} \qquad (6.4.1)$$

Each energy function gets the same terms tacked on,

$$dG = -S\,dT + V\,dP + \mu_1\,dN_1 + \mu_2\,dN_2 + \cdots \qquad (6.4.2)$$

As before (see Eq. (6.2.7)),

$$G = \mu_1 N_1 + \mu_2 N_2 + \cdots = \sum_i \mu_i N_i \qquad (6.4.3)$$

Problem 6.4
Find F and H for systems of two or more components.

7

Statistical mechanics for fixed and variable N

7.1 Statistical mechanics for fixed N

Let's start out with a review of what we know about statistical mechanics. Suppose we have a box, of fixed volume V, filled with N particles. For example it might be a box of perfect gas, in which case each state results in a list of occupation numbers for all the single-particle states.

If the box is isolated with energy U, then in equilibrium it has Γ possible equally probable states, and $S = k_B \log \Gamma$. If we know Γ for each U, we have solved the problem, but such a solution would be inconvenient (in fact it's completely impossible). Instead of specifying Γ or U let us specify T. We'll do it in the usual way: Say the system consists of a very large medium and a very small sample. The system as a whole has energy U_0 and volume V_0. The system is isolated so U_0 and V_0 are constants. Then in equilibrium the system has Γ_0 possible quantum states, and its entropy is $S_0 = k_B \log \Gamma_0$.

The medium consists of the difference between the system and the sample. If the sample has energy U and volume V, the medium has energy $U' = U_0 - U$ and volume $V' = V_0 - V$. The medium is always internally in equilibrium, so

$$dU' = T \, dS' - P \, dV' = T \, dS'$$

since $V' = V_0 - V$ is constant.

Now let us call one single quantum state of the sample α. That is, for the perfect gas, α stands for one particular list of 10^{23} single-particle states each occupied by one particle. A sample in state α has energy U_α. There are in total Γ_α states with the same energy. The state α is just one of these states.

The system, remember, has Γ_0 equally probable possible quantum states. The probability of finding the system in any one of its possible quantum states is $w_0 = 1/\Gamma_0$. Now suppose the sample is in the state α. Then, because the sample has energy U_α, the medium has energy $U_0 - U_\alpha$. Let the number of states of the medium with this much energy be Γ'_α.

Now, (the number of states of the system) = (the number of states of the sample) $\times$ (the number of states of the medium):

$$\Gamma = (1)\Gamma'_\alpha = \Gamma'_\alpha$$

Of the Γ_0 total possible states of the system, there are Γ'_α for which the sample is in the state α. Therefore, the random probability of finding the sample in the state α is

$$w_\alpha = \Gamma'_\alpha/\Gamma_0$$

The probability of finding the sample in any state of the system, w_0, is the same for all of its states. But w_α is not the same for all α because the larger U_α, the smaller $U' = U_0 - U_\alpha$. But smaller U' (the energy of the medium) means smaller Γ'_α (the number of choices) and therefore smaller w_α. So w_α gets small as U_α gets big. Remember, w_α is not the probability that the sample has energy U_α. It is the probability that the sample is in one particular quantum state with energy U_α. There will in general be many such states, Γ_α of them.

The thermodynamic average value of any property of the sample, f, is

$$\bar{f} = \sum_\alpha w_\alpha f_\alpha \tag{7.1.1}$$

where f_α is the value of f when the sample is in the state α (this is always what we mean by an average value).

Since the sample is always in some state,

$$\sum_\alpha w_\alpha = 1 \tag{7.1.2}$$

If we know all the w_α, these two equations permit us to compute anything. We can even compute the entropy of the medium this way, since it depends only on U_α and is therefore, in effect, a property of the sample:

$$S'_\alpha = k_B \log \Gamma'_\alpha$$

This is the entropy of the medium when the sample is in the state α. The thermodynamic average value of the entropy of the medium is

$$S' = \sum_\alpha w_\alpha S'_\alpha = \sum k_B w_\alpha \log \Gamma'_\alpha$$

and $S_0 = k_B \log \Gamma_0$, so

$$S_0 - S' = -\sum_\alpha k_B w_\alpha \log \Gamma'_\alpha + k_B \log \Gamma_0$$

$$= -k_B \sum_\alpha w_\alpha \log(\Gamma'_\alpha/\Gamma_0)$$

where we have used the trick

$$k_B \log \Gamma_0 = k_B \log \Gamma_0 \sum_\alpha w_\alpha = k_B \sum_\alpha w_\alpha \log \Gamma_0$$

But $\Gamma'_\alpha / \Gamma_0 = w_\alpha$ and $S_0 - S' = S$, so

$$S = -k_B \sum_\alpha w_\alpha \log w_\alpha \qquad (7.1.3)$$

This is a famous and important formula.

7.2 The Gibbs distribution and the partition function

We'll now derive an expression for w_α as a function of U_α. Recall that $S'_\alpha = k \log \Gamma'_\alpha$, meaning that the entropy of the medium when the sample is in the state α depends on the number of states of the medium in that circumstance. So

$$S_0 - S'_\alpha = -k_B \log(\Gamma'_\alpha/\Gamma_0)$$

$$= k_B \log w_\alpha$$

Or, in other words,

$$w_\alpha = e^{-(S_0 - S'_\alpha)/k_B} = A e^{S'_\alpha/k_B}$$

where A is a constant because S_0 is a constant.

S'_α is the entropy of the medium when it has energy $U' = U_0 - U_\alpha$,

$$S'_\alpha = S'(U_0 - U_\alpha)$$

But the sample is small. We assume that $U_\alpha \ll U_0$. Of course, we should allow states that violate this condition, but w_α will be negligibly small for such states. Thus we can write a Taylor series for S'_α of the form

$$S(x + \delta x) = S(x) + \frac{\partial S}{\partial x} \delta x$$

namely

$$S'_\alpha = S'(U_0) - \frac{\partial S'}{\partial U'} U_\alpha$$

The medium is always in equilibrium, so $dU' = T\, dS'$ or $\partial S'/\partial U' = 1/T$. The quantity $S'(U_0)$ is the entropy that the medium would have if it contained all of the energy of the system. It has no particular significance except that it is constant, meaning that it does not depend on α. So

$$S'_\alpha = \text{constant} - U_\alpha/T$$

or

$$w_\alpha = Ae^{S'_\alpha/k_B} = Be^{-U_\alpha/(k_B T)}$$

where B is a new constant determined by

$$\sum_\alpha w_\alpha = B\sum_\alpha e^{-U_\alpha/(k_B T)} = 1$$

We have, finally,

$$w_\alpha = Be^{-U_\alpha/(k_B T)} \qquad (7.2.1)$$

where

$$B = \frac{1}{\sum_\alpha e^{-U_\alpha/(k_B T)}} \qquad (7.2.2)$$

Equations (7.2.1) and (7.2.2) are known as the **Gibbs distribution**.

Now, since $U = \sum_\alpha U_\alpha w_\alpha$ and $\sum_\alpha w_\alpha = B\sum_\alpha e^{-U_\alpha/(k_B T)}$, we have, for a sample at constant T,

$$U = \frac{\sum_\alpha U_\alpha e^{-U_\alpha/(k_B T)}}{\sum_\alpha e^{-U_\alpha/(k_B T)}}$$

If we have solved the quantum mechanical problem of the sample, we know all the possible states, α, and for each of them we know the energy, U_α. The above formula, then, permits us to compute the energy the sample will have if it's held at constant temperature.

This, however, is $U(T, V, N)$, not the proper variables, which would be S, V and N. The entropy of the sample is given by Eq. (7.1.3):

$$S = -k_B \sum_\alpha w_\alpha \log w_\alpha$$

$$= -k_B \sum_\alpha w_\alpha \left(\log B - \frac{U_\alpha}{kT} \right) = -k_B \log B \left(\sum_\alpha w_\alpha \right) + \frac{1}{T} \sum_\alpha w_\alpha U_\alpha$$

$$= -k_B \log B + \frac{U}{T}$$

In other words, $k_B T \log B = U - TS = F$, the free energy, or

$$F = -k_B T \log \left(\sum_\alpha e^{-U_\alpha/(k_B T)} \right) \tag{7.2.3}$$

where the above equation follows from the fact that $B\sum_\alpha e^{-U_\alpha/(k_B T)} = 1$. Given a list of states, α, and energies U_α, Eq. (7.2.3) gives F as a function of T, V and N, which are the proper variables, and thus the problem is solved. Let

$$Z = \sum_\alpha e^{-U_\alpha/(k_B T)} \tag{7.2.4}$$

where Z is the **partition function**, and

$$F = -k_B T \log Z \tag{7.2.5}$$

Suppose U_α is a sum of independent energy terms, that is,

$$U_\alpha = U_m + U_n + U_l$$

Then

$$Z = \sum_{m,n,l} e^{-(U_m + U_n + U_l)/(k_B T)}$$

$$= \sum_m e^{-U_m/(k_B T)} \sum_n e^{-U_n/(k_B T)} \sum_l e^{-U_l/(k_B T)}$$

$$= Z_m Z_n Z_l$$

In other words, the partition function partitions!

Problem 7.1
Write the (translational) partition function for a gas of N non-interacting molecules in volume V at a reasonable (i.e. not too low) temperature.

Problem 7.2
The molecules in the previous problem can each vibrate with energies $\varepsilon_n = (n + 1/2)hv$, where $n = 0, 1, 2, \ldots$ Find the vibrational partition function.

Problem 7.3
The molecules in the above problems can rotate, and the rotational partition function is the same as the translational one, except that the V is replaced by 4π, the mass is replaced by the moment of inertia, I, and the power $3/2$ is replaced by 1 since there are only two rotational degrees of freedom. Write the rotational partition function.

Problem 7.4
Calculate the pressure.

Problem 7.5
Calculate the energy.

Problem 7.6
Calculate the heat capacity at constant volume.

7.3 Statistical mechanics for variable N

Let us consider a system just like the one we had in the previous three chapters, but, keeping V fixed, we'll let N vary. This is called an open system. If the sample is in a state α, the sample has energy and number of particles U_α and N_α, and the medium has a number of states Γ'. We still have $w_\alpha = \Gamma'_\alpha/\Gamma_0$, $S'_\alpha = k_B \log \Gamma'_\alpha$ and

$$S_0 - S'_\alpha = k_B \log \Gamma_0 - k_B \log \Gamma'_\alpha = -k_B \log w_\alpha$$

So $w_\alpha = A e^{S'_\alpha/k_B}$, but now $S'_\alpha = S'(U_0 - U_\alpha, N_0 - N_\alpha)$, where $U_\alpha \ll U_0$ and $N_\alpha \ll N_0$, so

$$S'_\alpha = S'(U_0, N_0) - \frac{\partial S'}{\partial U'} U_\alpha - \frac{\partial S'}{\partial N'} N_\alpha \qquad (7.3.1)$$

Here $S'(U_0, N_0)$, the entropy the medium would have if it contained all of the energy and particles, is simply a constant, and

$$dS' = \frac{dU'}{T} - \frac{\mu}{T} dN'$$

so

$$\frac{\partial S'}{\partial U'} = \frac{1}{T}$$

and

$$\frac{\partial S'}{\partial N'} = -\frac{\mu}{T}$$

Thus

$$S'_\alpha = \text{constant} - \frac{U_\alpha}{T} + \frac{\mu N_\alpha}{T}$$

Recalling that $w_\alpha = Ae^{S_\alpha/T}$ and $\sum_\alpha w_\alpha = 1$, we have,

$$w_\alpha = Ce^{-(U_\alpha - \mu N_\alpha)/(k_B T)} \tag{7.3.2}$$

$$C = \frac{1}{\sum_\alpha e^{-(U_\alpha - \mu N_\alpha)/(k_B T)}} \tag{7.3.3}$$

Equations (7.3.2) and (7.3.3) constitute the Gibbs distribution for variable N. Then, for example,

$$\langle U \rangle = \frac{\sum_\alpha U_\alpha e^{-(U_\alpha - \mu N_\alpha)/(k_B T)}}{\sum_\alpha e^{-(U_\alpha - \mu N_\alpha)/(k_B T)}} = \langle U(T, V, \mu) \rangle \tag{7.3.4}$$

and

$$\langle N \rangle = \frac{\sum_\alpha N_\alpha e^{-(U_\alpha - \mu N_\alpha)/(k_B T)}}{\sum_\alpha e^{-(U_\alpha - \mu N_\alpha)/(k_B T)}} = \langle N(T, V, \mu) \rangle \tag{7.3.5}$$

We also have

$$S = -k_B \sum_\alpha w_\alpha \log w_\alpha$$

$$= -k_B \sum_\alpha w_\alpha \left(\log C - \frac{U_\alpha - \mu N_\alpha}{k_B T} \right)$$

$$= -k_B \log C + \frac{U - \mu N}{T}$$

So

$$k_B T \log C = U - TS - \mu N = F - \mu N = \Omega$$

Or

$$\Omega = -k_B T \log \Xi \tag{7.3.6}$$

where

$$\Xi = \sum_\alpha e^{-(U_\alpha - \mu N_\alpha)/(k_B T)} \tag{7.3.7}$$

and Ξ is called the grand partition function.

That completes our survey of statistical mechanics for both fixed and variable N. In Chapter 8 we will turn to some more advanced topics

Problem 7.7
Write the grand partition function for an ideal gas with a variable number of particles.

8

More advanced topics

8.1 Interacting particles

Suppose we have a sample of particles (atoms or molecules) that do interact with one another. Let's say the system has a potential energy W, which in principle depends on all the positions of the particles. Then the total energy of the system can be written

$$U = \sum_{i=1}^{N} \frac{p_i^2}{2m} + W \tag{8.1.1}$$

or, in other words,

$$e^{-U/(k_{\mathrm{B}}T)} = e^{-\sum_i P_i^2/(2mk_{\mathrm{B}}T)} e^{-W/(k_{\mathrm{B}}T)}$$

where p is the momentum and m the mass of the particle. If this were for a single particle, we would have $W = 0$ and

$$U = (p_x^2 + p_y^2 + p_z^2)/(2m)$$

and the partition function would be

$$Z_1 = \sum_{n_x, n_y, n_z} e^{-U/(k_{\mathrm{B}}T)}$$

where $p_x = n_x p_0$ and so on for y and z. For N interacting particles we cannot simply restore the W and write $Z = Z_1^N$ since we would then be counting each state too many times. There would be a separate term for each single particle state for particle 1, with all others fixed, a term for each state of particle 2, with all the others fixed, and so on. In other words we would be counting each state $N!$ times (as long as the temperature is not very low). We can easily repair this difficulty by writing $Z = Z_1^N/N!$,

$$Z = \frac{1}{N!} \left(\sum_{i=1}^{3N} e^{-p_i^2/(2mk_{\mathrm{B}}T)} e^{-W/(k_{\mathrm{B}}T)} \right)^N \qquad (8.1.2)$$

Problem 8.1

Why in the above must the temperature not be too low?

8.2 Doing sums as integrals

For interacting particles, their energy, U, depends not only on $3N$ coordinates in p-space, but also on $3N$ coordinates in real space. Thus they exist in a $6N$-dimensional space known as ***phase space***. The number of possible states in the $6N$-dimensional phase space is $X^*(U)$,

$$X^*(U) = \frac{V^*(U)}{N! \tau_0^{6N}} \qquad (8.2.1)$$

where $V^*(U)$ is the $6N$-dimensional volume that the system takes up in phase space, $N!$ plays the same role as it plays in Eq. (8.1.2), and τ_0^6 is the minimum volume of a single state in phase space. Each state has a minimum value because of the uncertainty principle, $\Delta x\, \Delta p_x \geq h$, where h is Planck's constant.

We can figure out what τ_0 is because the general formula must work for a special case: a single particle in a box. The volume in phase space is the volume in ordinary space multiplied by the volume of a sphere in momentum space. We have

$$X_1^*(\varepsilon) = \frac{V(4/3)\pi p^3}{\tau_0^6}$$

so that

$$\rho(\varepsilon) = \frac{\partial X_1^*}{\partial \varepsilon} = \frac{V}{\tau_0^6} 4\pi p^2 \frac{\partial p}{\partial \varepsilon}$$

But we saw earlier that

$$\rho(\varepsilon) = \frac{4\pi p^2}{p_0^3} \frac{\partial p}{\partial \varepsilon}$$

or, in other words,

$$\tau_0^6 = p_0^3 V = (h/L)^3 V = h^3$$

This must be the same for all systems, $\tau_0^2 = h$. So

$$X_1^*(U) = V^*(U)/(N!h^{3N})$$

The interpretation of this is that each particle for each coordinate has some value of the coordinate and its momentum, say, x and p_x. We divide x into equal units, Δx, and p_x into equal units Δp_x such that

$$\Delta x\, \Delta p_x = h$$

According to the uncertainty principle, this is the best we can do. Then, for each state, rather than giving x and p_x, we say it's somewhere in $\Delta x\, \Delta p_x$. Then, the number of states of one particle in one dimension is $\int dp\, dx/h$. The number of states of N particles in three dimensions is

$$\frac{1}{N!} \int_U \frac{d^{3N}p\, d^{3N}x}{h^{3N}}$$

Our notation is

$$d^{3N}x\, d^{3N}p = dx_1\, dy_1\, dz_1\, dx_2\, dy_2\, dz_2 \ldots$$
$$dx_N\, dy_N\, dz_N\, dp_{x1}\, dp_{y1}\, dp_{z1} \ldots dp_{xN}\, dp_{yN}\, dp_{zN}$$

over $6N$ Cartesian coordinates. If the system is isotropic,

$$dx\, dy\, dz = 4\pi r^2\, dr$$

and

$$d^3p = 4\pi p^2\, dp$$

In any case,

$$Z = \int \rho(U)e^{-U/(k_BT)}\, dU$$

$$= \int \frac{dX^*}{dU} e^{-U/(k_BT)}\, dU = \int e^{-U/(k_BT)}\, dX^*$$

$$= \frac{1}{N!} \int \frac{dV^*}{h^3} e^{-U/(k_BT)}$$

Or, finally,

$$Z = \frac{1}{N!} \int e^{-U/(k_BT)} \frac{d^{3N}x\, d^{3N}p}{h^{3N}} \tag{8.2.2}$$

Problem 8.2

Find the partition function for a single anharmonic oscillator in three dimensions whose energy is

$$\varepsilon = \frac{p^2}{2m} + br^{2n}$$

for $n > 1$ and $r^2 = x_1^2 + x_2^2 + x_3^2$.

8.3 The partition function for interacting particles

It's time for some fancy arithmetic. We'll start with

$$U = \sum_{i=1}^{3N} \frac{p_i^2}{2m} + W$$

where the potential energy W depends on all of the x_i. Then

$$Z = \frac{1}{N!} \int \exp\left(-\sum_{i=1}^{3N} \frac{p_i^2}{2mk_BT}\right) d^{3N}p_i \int e^{-W/(k_BT)} \frac{d^{3N}x}{h^{3N}}$$

$$= \frac{1}{N!h^{3N}} \left[\int \exp\left(-\frac{p^2}{2mk_BT}\right) d^3p\right]^N \int e^{-W/(k_BT)} d^{3N}x$$

Let

$$Q_N = \int e^{-W/(k_BT)} d^{3N}x \qquad (8.3.1)$$

This is called the ***configurational integral***. It depends only on T and V. The partition function becomes

$$Z = \frac{Q_N}{N!h^{3N}} \left[\int_0^{\infty} e^{-p^2/(2mk_BT)} 4\pi p^2 \, dp\right]^N \qquad (8.3.2)$$

$$= Q_N Z_1^N / (V^N N!)$$

where Z_1 is the partition function for a single particle. For an ideal gas,

$$Z_{IG} = \frac{Z_1^N}{N!} \qquad (8.3.3)$$

And, finally,

$$Z = Z_{IG} \frac{Q_N}{V^N} \qquad (8.3.4)$$

is the partition function for any classical system.

The system is classical because we've specified that the temperature not be very low. We can put that in different terms as follows. From Eq. (5.4.22),

$$Z_1 = (k_BT)^{3/2} \frac{2\sqrt{2\pi^3}}{h^3} V m^{3/2}$$

So, we can write

$$Z_1 = V/\Lambda^3$$

where $\Lambda = h/\sqrt{2\pi m k_B T}$.

Then our criterion that the temperature not be too low can be written

$$N\Lambda^3/V \ll 1$$

8.4 What happened?

Let us review very briefly what we've just seen.

8.4.1 The density of states

Consider a single particle with momentum between zero and p, at a position anywhere in the volume V. Then the allowed volume in p-space times the allowed volume in real space is

$$X(p) = \frac{4\pi p^3}{3h^3} V$$

and the density of states is

$$\rho(\varepsilon) = \frac{dX}{d\varepsilon}$$

or

$$\rho(\varepsilon)d\varepsilon = dX = \frac{d^3p\, d^3x}{h^3}$$

If we have N particles, this applies to each one independently if they are distinguishable, but they are indistinguishable, so we take the product and divide by $N!$:

$$\rho(U)dU = \frac{1}{N!} \frac{d^{3N}p\, d^{3N}x}{h^{3N}}$$

We already had this for the case of an ideal gas. For the case of an interacting system,

$$U = \sum_i \frac{p_i^2}{2m} + W \qquad (8.4.1)$$

8.4.2 The partition function

The partition function is given by

$$Z = \int \rho(U) e^{-U/(k_B T)} \, dU = \frac{1}{N!} \int \cdots \int e^{-U/(k_B T)} \frac{d^{3N}p \, d^{3N}x}{h^{3N}}$$

After substituting in Eq. (8.4.1) and a little rearranging, this gives

$$Z = \frac{Z_{IG}}{V^N} Q_N$$

where

$$Q_N = \int \cdots \int e^{-W/(k_B T)} \, d^{3N}x$$

This describes everything that happens in nature except for certain phenomena at low temperature, where the indistinguishability of states can't be described simply by dividing by $N!$ and where sums cannot be safely done as integrals. Since we've already solved the ideal gas part completely, it all comes down to finding Q_N. For example,

$$F = -k_B T \log Z$$
$$= -k_B T \log Z_{IG} - k_B T \log\left(\frac{Q_n}{V^N}\right)$$

Then

$$P = -(\partial F/\partial V)_T$$
$$= N k_B T/V + k_B T \left[\frac{1}{Q_N}\left(\frac{\partial Q_N}{\partial V}\right)_T - \frac{N}{V}\right]$$

Or

$$P = \frac{k_B T}{Q_N}\left(\frac{\partial Q_N}{\partial V}\right)_T \qquad (8.4.2)$$

Then, if the system really is an ideal gas, $W = 0$ and

$$Q_N = \int e^{-W/(k_B T)} \, d^{3N}x = V^N$$

so that $P = Nk_BT/V$, as it should. In a real system, we need to know its potential energy as a function of where the atoms are. Notice that, in this part of the calculation, the atoms are to be regarded as distinguishable.

8.5 The statistical mechanics of real molecules

For an ideal gas,

$$U = \frac{p_1^2}{2m} + \cdots + \frac{p_N^2}{2m}$$

For a real gas (or any other system)

$$U = \frac{p_1^2}{2m} + \cdots + \frac{p_N^2}{2m} + W(\vec{x_1}, \ldots, \vec{x_N})$$

Our problem is to compute Z, where $Z = \int \rho(U)e^{-U/(k_BT)} \, dU$. We need to find $\rho(U)$. For the ideal gas we had

$$\text{number of states up to } p = X = \frac{4}{3}\pi p^3 / p_0^3$$

where p_0 is the volume element in phase space, $p_0 = h$. Then the density of states is

$$\rho(\varepsilon) = \frac{\partial X}{\partial p}\frac{\partial p}{\partial \varepsilon} = \frac{4\pi p^2}{p_0^3}$$

For many atoms, we have the same argument, leading to

$$X(U) = \frac{\text{volume}}{N! p_0^{3N}}$$

where the volume is the volume of a $3N$-dimensional momentum space. If the particles interact, we will have to evaluate

$$Q_N = \iint e^{-W/(k_BT)} \, d^{3N}x$$

To evaluate this integral, we must know the form of $W(N)$.

8.6 Two atoms

Start with two atoms. Two neutral atoms have zero potential, except for their mutually induced dipole moments. The dipole moment of an atom in an electric field is given by $\vec{p} = \alpha \vec{E}$, where α is called the polarizability. The energy of a dipole in an electric field is

$$u = -\vec{p} \cdot \vec{E} = -\alpha E^2$$

The electric field due to the other dipole is $E \propto 1/r^3$, so the energy due to the induced dipole is $u \propto -1/r^6$, which is attractive at large distances. At short distances the electron clouds start to overlap and the potential becomes strongly repulsive. There is no exact form, but an r^{-12} potential does nicely. The result is

$$u(r) = 4\varepsilon_0 \left[\left(\frac{\sigma}{r} \right)^{12} - \left(\frac{\sigma}{r} \right)^6 \right] \tag{8.6.1}$$

Now consider a third atom. It has the same effect on each of the other two atoms, but their interaction doesn't affect the third. This is called pairwise additivity:

$$W = u(r_{12}) + u(r_{13}) + u(r_{23})$$

If there are N atoms,

$$W = \sum_{i<j}^{N} u(r_{ij})$$

where $i < j$ means we don't count separately u_{12} and u_{21}. We could equally well have written

$$W = \frac{1}{2} \sum_{i,j}^{N} u(r_{ij})$$

Then, the configurational integral becomes

$$Q_N = \iiint \exp\left(-\sum_{i,j} \frac{u(r_{ij})}{2k_\mathrm{B}T} \right) d^{3N} r_N$$

For example, if only pairs of atoms are close enough together to have a nonzero $u(r)$,

$$Q_2 = \iint \exp\left(-\frac{1}{2k_\mathrm{B}T} [u(r_{12}) + u(r_{21})] \right)$$

$$= \iint e^{-u(r_{12})/(k_\mathrm{B}T)} \, d^3 r_1 \, d^3 r_2$$

The two integrals are to be done separately over the volume V. The notation means

$$d^3r_1 = dx_1 \, dy_1 \, dz_1 \text{ (Cartesian coordinates)}$$
$$= r_1^2 \sin^{-1}\theta_1 \, d\Theta_1 \, d\Phi_1 \, dr_1 \text{ (spherical coordinates)}$$
$$= 4\pi r_1^2 \, dr_1 \text{ (isotropic)}$$

In

$$Q_2 = \iint e^{-u(r_{21})} \, d^3r_1 \, d^3r_2$$

we hold 1 fixed while we integrate over r_2, then we integrate over r_1. But the exponent depends only on

$$r_{12} = r_1 - r_2$$

and we have

$$dr_{12} = dr_1 - dr_2$$

In the integral over $\vec{r}_1$, $\vec{r}_2$ is fixed, $d\vec{r}_2 = 0$ and $dr_{12} = dr_1$. The net result is that

$$Q_2 = \iint e^{-u(r_{12})/(k_B T)} \, d^3r_{12} \, d^3r_2$$

Now the integrand in this expression doesn't depend on r_2, so that integration just gives a factor V, yielding

$$Q_2 = V \int e^{-u(r_{12})/(k_B T)} \, d^3r_{12}$$

But now r_{12} is a dummy variable, and the integration is isotropic:

$$Q_2 = 4\pi V \int e^{-u(r)/(k_B T)} r^2 \, dr \qquad (8.6.2)$$

and we've reduced the whole thing to a problem of integration.

8.7 More atoms

We have

$$Q_N = \iint \exp\left(-\sum_{i<j} \frac{u(r_{ij})}{k_B T}\right) d^{3N} x_i$$

which is the same as

$$Q_N = \iint \prod_{i<j} e^{-u(r_{ij})/(k_B T)} \, d^{3N} x_i$$

Joseph Mayer and Maria Goeppert Mayer

There is a simple trick for handling this integral due to Joseph Mayer (a theoretical chemist and the husband of famous physicist Maria Goeppert Mayer). We write

$$f_{ij} = e^{-u(r_{ij})/k_B T} - 1$$

Then for $r_{ij} \to \infty$, $e^{-u(r_{ij})} \to 1$ and $f_{ij} \to 0$. For $r_{ij} \to 0$, $e^{-u(r_{ij})/(k_B T)} \to 0$ and $f_{ij} \to -1$. In general,

$$Q_N = \iint \prod_{i<j} (1 + f_{ij}) d^{3N} x_i$$

Let us multiply out the integrand,

$$\begin{aligned}
\prod_{i<j}(1 + f_{ij}) &= (1 + f_{12})(1 + f_{13}) \cdots (1 + f_{1N})(1 + f_{23}) \cdots \\
&= 1 + f_{12} + f_{13} + \cdots + f_{1N} + f_{23} + \cdots \\
&\quad + f_{12} f_{23} + \cdots \\
&\quad + f_{12} f_{23} f_{13} + \cdots
\end{aligned} \tag{8.7.1}$$

and so on. But each $f_{ij} = 0$ unless i and j are close together. Furthermore, $f_{12} f_{23} = 0$ unless 1, 2 and 3 are all close together. This allows us to sort out the terms that are important. Products of fs are called ***clusters***, and integrals over clusters are called ***cluster integrals.***

The general idea is this: for a rarified gas, all the fs are zero, $\prod_{i<j}(1 + f_{ij}) = 1$, and we recover the ideal gas result. The first-order correction to that is the ***dilute gas***, in which two-particle collisions occur, but the probability of a three-particle collision, or of two two-particle collisions at the same time, is negligible. Then Eq. (8.7.1) becomes $\prod_{i<j}(1 + f_{ij}) = f_{12} + \cdots + f_{(N-1),N}$, but we ignore all products. At higher density we get more terms up to the full problem, the problem of classical liquids, which is still unsolved.

Let's look at the first few orders of approximation. For the ideal gas,

$$Q_N^{IG} = \iint d^{3N}x = V^N$$

or

$$Z = Z_{IG}Q_N^{IG}/V^N = Z_{IG}$$

No surprise there. For a dilute gas, we have

$$Q_N^{DG} = \iint (1 + f_{12} + \cdots + f_{N-1,N})d^{3N}x$$

$$= V^N + \iint f_{12}\,d^{3N}x + \cdots + \iint f_{N-1,N}\,d^{3N}x$$

All these integrals are the same,

$$\iint f_{ij}\,d^3x_1 \ldots d^3x_i \ldots d^3x_j \ldots d^3x_N$$

The integrand depends only on $|r_i - r_j|$, not on the others, so this gives

$$V^{N-2}\iint f_{ij}\,d^3x_i\,d^3x_j$$

It doesn't matter which two i and j are; all the integrals are the same as

$$V^{N-2}\iint f_{12}\,d^3x_1\,d^3x_2$$

Now use a trick: integrate over x_1, giving a factor V, then keep x_1 fixed so that f_{12} depends only on x_2. Then the above integral is given by

$$V^{N-2}V\int f_{12}\,d^3r_{12} = V^{N-1}\int 4\pi r^2 f(r)dr$$

where r is now a dummy variable. In the sum $f_{12} + \cdots + f_{N-1,N}$ there are $N(N-1)/2$ terms, so we have finally

$$Q_N^{DG} = V^N + \frac{N(N-1)}{2}V^{N-1}4\pi\int f(r)r^2\,dr \tag{8.7.2}$$

and

$$Z = Z_{IG}\frac{Q_N}{V^N} = Z_{IG}\left[1 + \frac{N(N-1)}{2V}4\pi\int_0^\infty f(r)r^2\,dr\right] \tag{8.7.3}$$

and the dilute gas problem is solved if we can evaluate the integral. Notice that $f(r) \to 0$ for large r; the integral should be evaluated only in V, but we can let it go to infinity. We have $f(r)$ depending only on T and the parameters of the potential, σ and ε_0. Let

$$B(T) = -\frac{4\pi}{2} \int f(r) r^2 \, dr \qquad (8.7.4)$$

Then

$$Z = Z_{IG} \left[1 - \frac{N(N-1)}{V} B(T) \right]$$

and we have

$$P = \left(\frac{\partial F}{\partial V} \right)_T$$

$$= -\frac{\partial}{\partial V}(-k_B T \log Z) = \frac{k_B T}{Z} \frac{\partial Z}{\partial V}$$

$$= \frac{k_B T}{Z_{IG}} \frac{\partial Z_{IG}}{\partial V} + \frac{k_B T (N/V)^2 B}{1 - N^2 B/V}$$

where we have taken $N(N-1) \approx N^2$, and for the dilute gas, $(N/V)^2 \ll N/V$, or $(N/V)^2 B \ll 1$, so

$$k_B T (N/V)^2 B / (1 - N^2 B/V) \approx k_B T (N/V)^2 B [1 + N^2 B/V]$$

$$= k_B T (N/V)^2 B + O\left[(N/V)^4 \right]$$

and

$$\frac{k_B T}{Z_{IG}} \frac{\partial Z_{IG}}{\partial V} = \frac{N k_B T}{V} = P_{IG}$$

so that

$$P = \frac{N k_B T}{V} + k_B T \left(\frac{N}{V} \right)^2 B$$

or, finally,

$$P = \frac{N k_B T}{V} \left[1 + \left(\frac{N}{V} \right) B(T) \right] \qquad (8.7.5)$$

This is actually the leading order of what is known as the **virial equation of state**.

8.8 More about equations of state

The simplest equation of state is the one for an ideal gas,

$$pV = Nk_BT$$

A slightly more complicated one, called the virial equation of state, includes higher-order terms. As we saw earlier, the full virial equation of state has the form

$$P = \frac{Nk_BT}{V}\left[1 + \frac{N}{V}B(T) + \left(\frac{N}{V}\right)^2 C(T) + \cdots\right] \tag{8.8.1}$$

where the first term in brackets refers to a gas of independent single particles, the second term includes two-particle interactions (as we have seen), the third term includes three-particle interactions and so on. C, D and so on have been calculated using the f_{ij} formalism up to fifth order, but the calculation is very tedious, and we won't bother with it beyond second order. We have found the second virial coefficient, which is given by

$$B(T) = -\frac{4\pi}{2}\int f(r)r^2\, dr$$

where $f(r) = e^{-u(r)/(k_BT)} - 1$ and the general form of $u(r)$ is shown in Fig. 8.1, in which the meanings of σ and u_0 are indicated.

Thus, for $r < \sigma$, $u(r) \to \infty$ rapidly and $e^{-u/(k_BT)} \to 0$, so $f(r) \approx -1$. For $r > \sigma$, $u(r)$ is always negative and never larger than ε_0. Let $k_BT \gg \varepsilon_0$. This condition at low density is where the virial equation works best. At much lower temperature, the gas may condense. If $k_BT \gg \varepsilon_0$, $e^{-u/(k_BT)} \approx 1 - u/(k_BT)$ and $f(r) \approx -u/(k_BT)$. Then

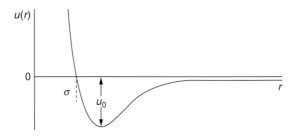

Figure 8.1 The interatomic potential.

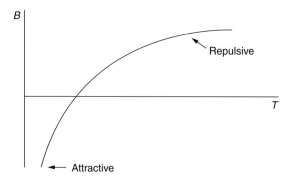

Figure 8.2 The behavior of $B(T)$.

$$B(T) = -\frac{4\pi}{2}\int_0^{\sigma} f(r)r^2\,dr + \frac{4\pi}{2}\int_{\sigma}^{\infty}\left(\frac{u}{k_\mathrm{B}T}\right)r^2\,dr$$

$$= b - \frac{a}{k_\mathrm{B}T}$$

where a and b are positive constants. So, $B(T)$ looks like Fig. 8.2.

Problem 8.3
Carry out the above calculation to the next order in N/V.

Another equation that represents the behavior of "real" gases is the ***van der Waals equation of state***,

$$\left[P + a\left(\frac{N}{V}\right)^2\right](V - Nb) = Nk_\mathrm{B}T \qquad (8.8.2)$$

This equation looks rather similar to the leading-order virial equation, but the relation between them can't be derived. It can only be arrived at by plausibility arguments. The leading-order virial is

$$PV = Nk_\mathrm{B}T\left(1 + \frac{N}{V}B\right) \qquad (8.8.3)$$

where $B = b - a/(k_\mathrm{B}T)$. First consider the hard core only. Then $a = 0$, so $B = b$. The van der Waals equation becomes

$$P(V - Nb) = Nk_\mathrm{B}T$$

and the virial equation becomes, $PV = Nk_\mathrm{B}T(1 + (N/V)b)$, or

$$\frac{PV}{1 + (N/V)b} = Nk_\mathrm{B}T$$

But, since $(N/V)b \ll 1$, we can write,

$$\frac{PV}{1 + (N/V)b} \approx PV\left(1 - \frac{N}{V}b\right) = P(V - Nb)$$

so the virial equation also becomes

$$P(V - Nb) = Nk_{\mathrm{B}}T$$

So the two equations become the same in this limit, which is the limit of high temperature and low density. If the temperature is high enough, this ought to work up to $V \approx Nb$, the density of the liquid. However, that's true only at high temperature or if the particles have hard-core potentials only.

Now we consider the attractive part only. We take $b = 0$, but a remains finite. This approximation ought to work at low density and high temperature. The virial equation becomes

$$PV = Nk_{\mathrm{B}}T\left(1 - \frac{aN}{k_{\mathrm{B}}TV}\right)$$

Or

$$P = \frac{Nk_{\mathrm{B}}T}{V} - a\left(\frac{N}{V}\right)^2$$

or, finally,

$$\left[P + a\left(\frac{N}{V}\right)^2\right]V = Nk_{\mathrm{B}}T$$

which is the van der Waals equation in this limit.

However, this ought to work only at low density. We'll use a different argument now, which is also valid at low density, to see why the van der Waals equation is not exactly right.

To study the attractive part of the potential, we'll use Eq. (8.4.2)

$$P = \frac{k_{\mathrm{B}}T}{Q_N}\frac{\partial Q_N}{\partial V}$$

where, of course,

$$Q_N = \iint e^{-W/(k_{\mathrm{B}}T)}\, d[N]$$

where $d[N] = d^{3N}x$. Now, instead of trying to find the form of $W[N]$, we'll approximate it as a constant – that is, we'll replace it by its average value before evaluating the integral.

$$Q_N \approx \iint e^{-\overline{W/(k_\mathrm{B}T)}} \, d[N] = V^N e^{-\overline{W/(k_\mathrm{B}T)}}$$

Recall that

$$W = \frac{1}{2} \sum_{i,j} u(r_{i,j})$$

On the average, all these terms are the same,

$$\overline{W} = \frac{1}{2} N \sum_{i,j} \overline{u(r_{i,j})} = \frac{1}{2} N^2 u_0$$

and

$$\overline{u(r_{i,j})} = u_0 = \frac{1}{V} \int u(r) d^3 r$$

is the average potential due to all the other atoms spread uniformly throughout the volume. Recall that

$$B(T) = -\frac{4\pi}{2} \int f(r) r^2 \, dr = -\frac{1}{2} \int f(r) d^3 r$$

At high temperature and low density, $e^{-u/(k_\mathrm{B}T)} \approx 1 - u/(k_\mathrm{B}T)$, so $f(r) = -u/(k_\mathrm{B}T)$. Then the attractive part of $B(T)$ is given by

$$B(T) = -\frac{1}{2} \int_\sigma^\infty (-u/(k_\mathrm{B}T)) d^3 r$$
$$= -a/(k_\mathrm{B}T)$$

so

$$u_0 = -2a/V$$

and

$$\overline{W} = -N^2 a/V$$

so

$$Q_N = \iint e^{-\overline{W/(k_\mathrm{B}T)}} \, d[N] = \iint e^{N^2 a/(k_\mathrm{B}TV)} \, d[N]$$
$$= V^N e^{N^2 a/(kT_\mathrm{B}V)}$$

and

$$\frac{\partial Q_N}{\partial V} = \left[NV^{N-1} - \left(\frac{N}{V}\right)^2 \frac{a}{k_B T} V^N\right] e^{N^2 a/(k_B T V)}$$

$$= \left[\frac{N}{V} - \left(\frac{N}{V}\right)^2 \frac{a}{k_B T}\right] Q_N$$

so that

$$P = \frac{k_B T}{Q_N} \frac{\partial Q_N}{\partial V} = k_B T \frac{N}{V} - a\left(\frac{N}{V}\right)^2$$

or, finally,

$$\left[P + a\left(\frac{N}{V}\right)^2\right] V = N k_B T$$

So, to get the attractive part of the potential, we replace P in the ideal gas equation by $P + a(N/V)^2$, which gives us the van der Waals equation in this limit.

Problem 8.4
For the attractive part of the potential, what is wrong with this procedure? Why can't we average over Q_N before doing the integral?

8.9 The van der Waals equation of state

As we've already seen, the van der Waals equation of state is

$$\left[P + a\left(\frac{N}{V}\right)^2\right](V - Nb) = N k_B T \qquad (8.9.1)$$

A plot of this equation looks like Fig. 8.3.

The equation gives a remarkably good account of the liquid and gas phases of a given substance, once it's properly interpreted. We'll interpret it now.

This equation is cubic in V. For each value of P, there are three solutions for V. Above a certain temperature, called the critical temperature or T_C, two of the solutions are imaginary and only one is real. Below T_C the solution goes through a loop so that one of the three solutions occurs where $(\partial P/\partial V)r$ is positive. That's unstable (a small increase in P causes V to

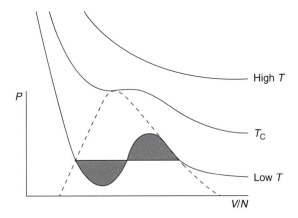

Figure 8.3 A plot of the van der Waals equation of state.

increase, which causes P to increase further and so on), so it can't represent a real point on the phase diagram. What happens instead is as follows.

At a given pressure, below where it starts to loop, a phase transition in which gas and liquid are in equilibrium occurs. At the phase transition the two chemical potentials are equal,

$$\mu_1 - \mu_g = 0 = \int_g^l \frac{V}{N} \, dp$$

where l and g subscripts mean liquid and gas. This gives the equal area construction shown in Fig. 8.3.

The critical point is the point at which the two phases merge into one. To find the critical point itself, we can use three equations, namely Eq. (8.9.1), plus $(\partial p/\partial V)_{T_C} = 0$ and $(\partial^2 p/\partial V^2)_{T_C} = 0$. These together give three equations for three unknowns, p_C, V_C and T_C. It may be easier to get the three values by solving $(V - V_C)^3 = 0$.

Problem 8.5
Write the van der Waals equation in reduced form, with $p = P/P_C$, $v = V/V_C$ and $t = T/T_C$.

Problem 8.6
Find V_C, p_C and $k_B T_C$. The results are $V_C = 3Nb$, $p_C = a/(27b^2)$ and

$$k_B T_C = \frac{8a}{27b}$$

Table 8.1 *Data for argon and xenon*

	ε_0/k_B (K)	T_C (K)
Argon	119.8	151
Xenon	225.3	290

Problem 8.7

Given that

$$u = 4\varepsilon_0 \left[\left(\frac{\sigma}{r} \right)^{12} - \left(\frac{\sigma}{r} \right)^{6} \right]$$

(the van der Waals potential) find the relation between a and b and σ and ε_0.

Solutions.

$$a = (16/9)\varepsilon_0\sigma^3 \text{ and } b = (2/3)\pi\sigma^3$$

Problem 8.8

Using the solutions to Problem 8.7, find $k_B T_C$, V_C and P_C in terms of σ and ε_0.

One result of Problems 8.6 and 8.7 is that $k_B T_C \approx \varepsilon_0$. This is a truly remarkable result. It means that 10^{23} atoms acting with mutual potential energies will start to condense when $T \approx \varepsilon_0 k_B$. We can see how well this works by looking at a couple of examples in Table 8.1. So it's not exactly right, but it does work approximately.

Let us put the van der Waals equation of state in a different form, in which it will deal with all gases at once. We define $t = T/T_C$, $p = P/P_C$ and $v = V/V_C$. Then, since

$$t = \frac{k_B T}{8a/(27b)}$$

the equation becomes

$$\frac{27b}{a} \left[P + a \left(\frac{N}{V} \right)^{2} \right] \left(\frac{V}{N} - b \right) = 8t$$

But

$$\frac{27b}{a}P = \frac{1}{b}\frac{P}{P_C} = \frac{p}{b}$$

so

$$\left[\frac{p}{b} + 27b\left(\frac{N}{V}\right)^2\right]\left(\frac{V}{N} - b\right) = 8t$$

or

$$\left[p + 27b^2\left(\frac{N}{V}\right)^2\right]\left(\frac{V}{Nb} - 1\right) = 8t$$

But $Nb = V_C/3$, so

$$\left[p + 3\left(\frac{3}{v}\right)^2\right](3v - 1) = 8t \qquad (8.9.2)$$

This puts the equation in universal form. We also have

$$P_C V_C/(Nk_B T_C) = 3/8 \qquad (8.9.3)$$

Problem 8.9
Prove Eq. (8.9.3).

   This equation is known as *the law of corresponding states*. And that completes our survey of more advanced topics.

Solutions to the problems

By Michael Beverland and David Goodstein

Chapter 1

1.1

$$T = (\partial U/\partial S)_{V,N} = \left(\frac{N}{V}\right)^{2/3} \exp\left(\frac{S}{3Nk_B/2} - s_0\right)$$

$$S = \frac{2}{3}Nk_B \log T \left(\frac{V}{N}\right)^{2/3} + s_0 = S(T,V)$$

$$V/N = k_B T/P$$

$$S = \frac{2}{3}Nk_B \log\left[\left(\frac{k_B T}{P}\right)^{2/3} T\right] + s_0 = S(T,P)$$

1.2

$$(\partial U/\partial N)_{S,V} = \mu$$

1.3

$$
\begin{aligned}
B &= 0,   &\Gamma &= 1 \\
B &= 1,   &\Gamma &= 2^3 = 8 \\
B &= 2,   &\Gamma &= 3 \times 2^2 = 12 \\
B &= 4,   &\Gamma &= 3 \times 2 = 6 \\
B &= 25, &\Gamma &= 8 + 2 \times 2^2 = 16
\end{aligned}
$$

1.4 We want the number of values of $\left(n_x^2 + n_y^2 + n_z^2\right)$ for particle 1 plus the same for particle 2 that add up to B. Start with $B = 24$. One such set of numbers is (9, 9, 4) (1, 1, 0). We get a factor $8 \times 4 = 32$ from the fact that each of the nonzero numbers can be $+$ or $-$. There are $3 \times 3 = 9$ possible arrangements of the numbers. Thus this combination gives $32 \times 9 = 288$ possibilities. There are many other possible combinations, including (4, 4, 4) (4, 4, 4), which gets only $8 \times 8 = 64$ from the $+$ and $-$. When we add all the possibilities up the result is 4116. Here are the three results:

$$B = 24, \qquad \Gamma = 4116$$
$$B = 25, \qquad \Gamma = 3906$$
$$B = 26, \qquad \Gamma = 5040$$

1.5 Unlike for $B = 25$, in which case the n^2s had to add up to an odd number, for $B = 24$ and 26 double occupancy of a single level is allowed. That reduces slightly the total number of possibilities.

1.6 We have the equations

$$dU_1 = T \, dS_1 + \mu_1 \, dN_1$$
$$dU_2 = T \, dS_2 + \mu_2 \, dN_2$$

So

$$dU_1 + dU_2 = 0 = T \, dS + (\mu_1 - \mu_2)dN_1$$

Maximizing the entropy gives

$$\mu_1 = \mu_2$$

1.7 In this case we have not (yet) made use of the second derivative, so we're looking at either stable or unstable equilibrium.

1.8 We leave out a negligible number of states when we write $\Gamma = \Gamma_1 \Gamma_2 \ldots$

1.9

$$\Gamma = \left[\left(\frac{mk_{\mathrm{B}}T}{2\pi\hbar^2} \right)^{3/2} \frac{e^{5/2}}{N/V} \right]^N$$

so

$$\Gamma(310 \times 290)/\Gamma(300 \times 300) = (310 \times 290)^{3N/2}/(300 \times 300)^{3N/2}$$

where $N = 6 \times 10^{23}$. Then

$$\frac{\Gamma(310 \times 290)}{\Gamma(300 \times 300)} = \left(\frac{89\,900}{90\,000} \right)^{9 \times 10^{23}}$$

$$= (1 - 0.001)^{9 \times 10^{23}}$$

Chapter 2

2.1 $V = Nk_{\mathrm{B}}T/P$; $(\partial V/\partial P)_T = -Nk_{\mathrm{B}}T/P^2$; $(\partial V/\partial T)_P = Nk_{\mathrm{B}}/P$

      $P = Nk_{\mathrm{B}}T/V$; $(\partial P/\partial V)_T = -Nk_{\mathrm{B}}T/V^2$; $(\partial P/\partial T)_V = Nk_{\mathrm{B}}/V$

We have

$$[1 - (\partial P/\partial V)_T(\partial V/\partial P)_T]dP = [(\partial P/\partial T)_V + (\partial P/\partial V)_T(\partial V/\partial T)_P]dT$$

so

$$[1 + (Nk_BT/V^2)(-Nk_BT/P^2)] = 0$$

and

$$(\partial P/\partial T)_V = (\partial P/\partial V)_T(\partial V/\partial T)_P$$

so

$$Nk_B/V + (-Nk_BT/V^2)(Nk_B/P) = Nk_B/V + (-Nk_BT/V^2)(V/T) = 0$$

2.2

$$(\partial T/\partial V)_S = (\partial P/\partial S)_V$$

$$(\partial P/\partial S)_V = (\partial P/\partial T)_V/(\partial S/\partial T)_V$$

$$(\partial P/\partial T)_V = -(\partial P/\partial V)_T(\partial V/\partial T)_P$$

so

$$(\partial T/\partial V)_S = -(\partial V/\partial T)_P(\partial P/\partial V)_T/(\partial S/\partial T)_V$$

2.3

$$\left(\frac{\partial S}{\partial T}\right)_V \left(\frac{\partial V}{\partial S}\right)_T \left(\frac{\partial T}{\partial V}\right)_S = -1$$

$$\left(\frac{\partial U}{\partial T}\right)_V = T\left(\frac{\partial S}{\partial T}\right)_V = -T\left(\frac{\partial S}{\partial V}\right)_T \left(\frac{\partial V}{\partial T}\right)_S$$

Define $F = U - TS$, so that $dF = -S\,dT - P\,dV$. Then

$$\frac{\partial^2 F}{\partial V\,\partial T} = \frac{-\partial P}{\partial T} = \frac{-\partial S}{\partial V}$$

So

$$\left(\frac{\partial U}{\partial T}\right)_V = -T\left(\frac{\partial P}{\partial T}\right)\left(\frac{\partial V}{\partial T}\right)_S$$

2.4 This one is done in the main text of the book.

2.5 $U(S,V)$ given;

$$U = \frac{3}{2}Nk_B\left(\frac{N}{V}\right)^{2/3} \exp\left[\frac{S}{(3/2)Nk_B} - s_0\right]$$

Find C_V, C_P, K_T and β:

$$P = -(\partial U/\partial V)_S = (2/3)U/V$$

$$T = (\partial U/\partial S)_V = U/[(3/2)Nk_B]$$

$$PV = Nk_B T$$

$$\beta = \frac{1}{V}\left(\frac{\partial V}{\partial T}\right)_P = \frac{1}{T}$$

$$K_T = -\frac{1}{V}\left(\frac{\partial V}{\partial P}\right)_T = \frac{1}{P}$$

$$C_V = T\left(\frac{\partial S}{\partial T}\right)_V = \frac{3}{2}Nk_B$$

$$C_P = C_V + \frac{TV\beta^2}{K_T} = \frac{3}{2}Nk_B + \frac{PV}{T} = \frac{5}{2}Nk_B$$

2.6 We just had $C_P = (5/2)Nk_B$;

$$C_P = C_V + \frac{TV\beta^2}{K_T} = \frac{3}{2}Nk_B + \frac{TV(1/T)^2}{1/P} = \frac{3}{2}Nk_B + \frac{PV}{T} = \frac{5}{2}Nk_B$$

$$V = Nk_B T/P = \frac{Nk_B}{P}\left[\left(\frac{N}{V}\right)^{2/3}\exp\left[\frac{S}{(3/2)Nk_B} - s_0\right]\right]$$

$$V^{5/3} = \frac{N^{5/3}k_B}{P}\exp\left[\frac{S}{(3/2)Nk_B} - s_0\right]; \quad \left(\frac{\partial V}{\partial P}\right)_S = -\frac{3}{5}V/P; \quad K_S = 3/(5P)$$

$$K_S = \frac{C_V}{C_P}K_T = 3/(5P)$$

2.7 $C_V = aVT^3$. This result is valid for a solid, so V is essentially constant

$$dS = aVT^2\,dT$$
$$S = aVT^3/3$$
$$dU = T\,dS = aVT^3\,dT$$
$$U = aVT^4/4$$
$$P = (\partial U/\partial V)_S = aT^4/4$$

So, $PV/U = 1$ and the equation of state is $PV = aVT^4/4$.

2.8

$$C_P = C_V + TV\beta^2/K_T = C_V + TVK_T(\beta/K_T)^2$$

so

$$C_P = C_V + TVK_T(\partial P/\partial T)_V^2$$

2.9 Done in the main text of the book.

2.10

$$\left(\frac{\partial V}{\partial P}\right)_S = -\left(\frac{\partial V}{\partial S}\right)_P \left(\frac{\partial S}{\partial P}\right)_V = -\frac{C_V}{T}\left(\frac{\partial V}{\partial S}\right)_P \left(\frac{\partial T}{\partial P}\right)_V$$

$$= -\frac{C_V}{T}\frac{(\partial T/\partial P)_V}{(\partial S/\partial T)_P(\partial T/\partial V)_P} = -\frac{C_V}{C_P}\left(\frac{\partial T}{\partial P}\right)_V \left(\frac{\partial V}{\partial T}\right)_P$$

but

$$\left(\frac{\partial T}{\partial P}\right)_V \left(\frac{\partial V}{\partial T}\right)_P = -\left(\frac{\partial V}{\partial P}\right)_T$$

So

$$K_S = -\frac{1}{V}\left(\frac{\partial V}{\partial P}\right)_S = -\frac{C_V}{C_P}\frac{1}{V}\left(\frac{\partial V}{\partial P}\right)_T = \frac{C_V}{C_P}K_T$$

2.11 Consider an isolated box of ideal gas with a moveable partition so that $V = \text{constant} = V_1 + V_2$. Minimize S with respect to U_1:

$$\partial S/\partial U_1 = \partial S_1/\partial U_1 - \partial S_2/\partial U_2 = 1/T_1 - 1/T_2 = 0$$

Then

$$\partial^2 S/\partial U_1^2 = -1/(T_1^2 C_{V1}) - 1/(T_2^2 C_{V2}) \le 0$$

Let the two sides be equal: $C_V \ge 0$.

2.12 Let the box in problem 2.11 be in a temperature bath. Then

$$(\partial^2 U/\partial V_1^2)_T = \partial(P_1 - P_2)/\partial V_1 = -\partial P_1/\partial V_1 - \partial P_2/\partial V_2$$
$$= 1/(V_1 K_{T1}) + 1/(V_2 K_{T2}) \ge 0$$

Let the two sides be equal again, $K_T \ge 0$.

2.13

$$C_P = C_V + TVK_T\left(\frac{\partial P}{\partial T}\right)_V^2$$

All terms on the right-hand side are greater than zero; therefore, so is C_P. Also

$$K_S = \frac{C_V}{C_P} K_T$$

The same as above applies.

2.14 Done in the main text of the book.

2.15 We have

$$U = B \frac{h^2}{2mV^{2/3}}$$

As explained in the text, constant B means constant entropy. Thus

$$(\partial U/\partial V)_S = -P = -(2/3)U/V$$

2.16 For a free expansion, $\partial U/\partial V = 0$, $\partial T/\partial V = 0$ and $\partial S/\partial V = Nk_B/V$.

For an isothermal expansion, $\partial U/\partial V = 0$, $\partial T/\partial V = 0$ and $\partial S/\partial V = P/T$.

For an adiabatic expansion, $(\partial U/\partial V)_S = -P$, $(\partial T/\partial V)_S = -2P/(3Nk_B)$ and $\partial S/\partial V = 0$.

For an isobaric expansion, $(\partial U/\partial V)_P = 3P/2$, $(\partial T/\partial V)_P = P/(Nk_B)$ and $(\partial S/\partial V)_P = (5/3)P/T$.

2.17 For a free expansion the curve goes down from P_1V_1 to P_2V_2; PV = constant.

Isothermal expansion is the same.

Adiabatic expansion goes down more steeply: PV^γ = constant, with $\gamma = 5/3$.

For isobaric expansion the curve is a horizontal line.

2.18 For a free expansion, the work done is 0. The entropy change is $Nk_B \log 2$. Final $T = T_1$ and final $P = P_1/2$.

For an isothermal expansion, the work done is $Nk_BT \log 2$. The entropy change is $Nk_B \log 2$. $T = T_1$ and $P = P_1/2$.

For an adiabatic expansion, the work done is

$$\int_{V_1}^{2V_1} P \, dV = (3/2)Nk_BT(1 - 1/\sqrt{2})$$

The entropy change is 0. $T = 2PV_1/(Nk_B)$ and $P = P_1/2^\gamma$.

For an isobaric expansion, the work done is $2Nk_BT$. The entropy change is $(5/3)Nk_B \log 2$. $T = 2T_1$ and $P = P_1$.

2.19 $U_2 = U_1 +$ work done on the gas $= U_1 + P_1V_1 - P_2V_2$. So

$$U_1 + P_1V_1 = U_2 + P_2V_2$$

2.20 Ideal gas: $PV = Nk_BT$ and $U = (3/2)Nk_BT$, so

$$T_2 - T_1 = \frac{2}{3}\frac{U_1 - U_2}{Nk_B} = \frac{P_1V_1 - P_2V_2}{Nk_B}$$

Chapter 3

3.1 Triple point $T_3 = k_BT_k = 3.8 \times 10^{-14}$ erg. For NBP $T_3 = 5.15 \times 10^{-14}$ erg.

3.2 Zero degrees Kelvin $= -459.7$ degrees Fahrenheit. Fahrenheit and Celsius are the same at -40 degrees. Boltzmann's constant on the Rankine scale $= 7.67 \times 10^{-17}$ erg/°F.

3.3 $K_T = -(1/V)(\partial V/\partial P)_T$. A finite change in V with no change in P occurs at a phase transition. See Fig. P3.3.

3.4 Increase P on pure solid ice to melt at higher density. Decrease P on another solid to melt at lower density.

3.5 See Fig. P3.5.

3.6
$$C_P = (\partial U/\partial T)_P + P(\partial V/\partial T)_P = (7/2)Nk_B$$
$$C_V = (5/2)Nk_B$$
$$\gamma = C_P/C_V = 7/5$$

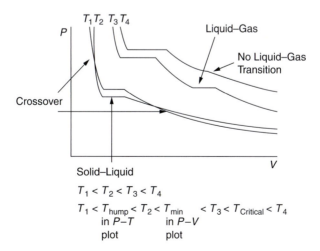

Figure P3.3

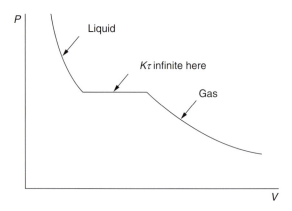

Figure P3.5

3.7

$$K_T = \frac{V}{Nk_BT(1 + 2BN/V)}$$

$$\beta = (1/T)\frac{1 + BN/V + (dB/dT)NT/V}{1 + 2BN/T}$$

3.8

$$K_T = (1 - Nb/V)/\left[(P - a(N/V)^2)(1 + 2Nb/V)\right]$$

$$\beta = Nk_B/[a(N/V)^2(2Nb - V) + PV]$$

3.9 Answers given. $C \approx 1$ because dimensional argument ought to work reasonably well.

3.10

$$P \approx (Nk_BT/V)[1 + (N/V)(b - a/(k_BT))] \approx (Nk_BT/V)(1 + NB/V)$$

$B = b - a/(k_BT)$, so, for high T, B is positive; and for low T, B is negative.

3.11 From the virial equation,

$$\frac{N}{V} = \frac{1}{2B_0}\left[-1 + \left(1 + \frac{4B_0P_0}{k_BT_0}\right)^{1/2}\right] \approx 2.68 \times 10^{15} \text{ particles/cubic meter}$$

whereas $(N/V)(\text{ideal}) = 1.73 \times 10^{15}$ particles/cubic meter. So the percentage error is 35%.

3.12

$$\sigma = \left(\frac{3b}{2\pi}\right)^{1/3} = 4.1 \times 10^{-9} \text{ m}$$

3.13

$$C_V = \frac{3}{2}Nk_B$$

$$C_P = C_V + TV\beta^2/K_T = \frac{3}{2}Nk_B + \frac{Nk_B(1+Nb/V)^2}{1+2NB/V}$$

$$U = (3/2)Nk_BT - aN^2/V$$

as given.

3.14 The pressure is 1 atm at sea level.

3.15 At higher altitude the pressure is lower than 1 atm and water boils at lower temperature. A pressure cooker is needed to make water boil at $\geq 100\,°C$.

3.16

$$P(h) = P_0 e^{-mgh/(k_BT)} \simeq P_0[1 - mgh/(k_BT)]$$

$$h \geq 20k_BT\,\Delta P/(mgP_0) \simeq 100\,\text{m}$$

3.17

$$P(h) = P(0)\left[1 - \frac{(\gamma-1)mg}{\gamma k_BT(0)}h\right]$$

$$h \ll \gamma k_BT(0)/(\gamma-1)mg \simeq 1\,\text{km}$$

Chapter 4

4.1 In step 1–2 $W_{12} = Nk_BT_H \log(V_2/V_1) = Q_{12}$; in step 2–3

$$W_{23} = \frac{Nk_B(T_H - T_L)}{\gamma - 1}$$

but $Q_{23} = 0$, so

$$W_{12} + W_{23} = Nk_BT_H \log(V_2/V_1) + Nk_B(T_H - T_L)/(\gamma - 1)$$

$$Q_{12} + Q_{23} = Nk_BT_H \log(V_2/V_1)$$

4.2 $S = S(T, V)$, so $T = T(S, V)$ and $T(S_1,V_1) = T_H$.

4.3 $V_4 = V_1(T_H/T_L)^{1/(\gamma - 1)}$.

4.4 Given Problems 4.1 and 4.3,

$$\oint P\,dV = (T_H - T_L)Nk_B \log(V_2/V_1)$$

4.5 $\oint P\,dV$ and $\oint V\,dP$ are both equal to the area enclosed by the same loop on a P–V diagram.

4.6 $S_2 - S_1 = Q_{12}/T_H = Nk_B \log(V_2/V_1)$, so

$$\oint P \, dV = (T_H - T_L)(S_2 - S_1)$$

4.7 $\eta = 1 - 300/2000 = 0.85$.

4.8

$$V_2 = (T_0/T_1)^{1/(\gamma-1)} V_1$$
$$V_3 = (T_0/T_1)^{1/(\gamma-1)} V_0$$
$$\eta = 1 - r^{\gamma-1}$$

Let R_A be Avogadro's number. Then

$$t_0 + t_2 = R_A \left(\frac{T_0}{T_H - T_0} + \frac{T_L}{T_1 - T_L} \right) R \log(V_1/V_0)$$

The work output is

$$W/t = \frac{(T_0 - T_1)(T_H - T_0)(T_1 - T_L)}{(T_1 T_H - T_0 T_L) R}$$

4.9 No, $Q \neq T \, dS$ because the free expansion is not reversible. No work is done, $\Delta W = 0$. $\Delta U = 0$, $W = 0$ and $Q = 0$.

4.10 $(\partial P/\partial V)_T = -Nk_B T/V^2 < 0$ and $(\partial P/\partial V)_S = -\gamma P/V < 0$.
    For the van der Waals gas

$$\left(\frac{\partial P}{\partial V} \right)_T = \frac{2a}{V} \left(\frac{N}{V} \right)^2 - \frac{Nk_B T}{V - Nb}$$

can be positive or negative.

4.11 At the four corners of the $P-V$ plot, $T_1 = P_1 V_1/R_A$, where R_A is Avogadro's number, $T_2 = P_1 V_2/R_A$, $T_3 = P_2 V_2/R_A$ and $T_4 = P_2 V_1/R_A$.

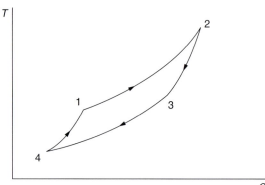

Figure P4.11

The *T–S* diagram is shown in Fig. P4.11.

It needs many temperature baths to be reversible. If it uses one high-temperature and one-low temperature bath it is less efficient than a Carnot cycle.

4.12 No work is done and no heat flows, so $X = P_2V_2 - P_1V_1$.

4.13 We want $H(S, P)$. Start from

$$U = \frac{3}{2}N\left(\frac{N}{V}\right)^{2/3}\exp\left[\frac{S}{(3/2)Nk_B} - \frac{s_0}{k_B}\right]$$

and $H = U + PV$. We need to use $V = V(S, P)$. It turns out that

$$V = NP^{-3/2}\exp\left[\frac{(2/5)S - (3/5)Ns_0}{Nk_B}\right]$$

so the result is

$$H(S, P) = \frac{5}{2}NP^{2/5}\exp\left[\frac{S}{(5/2)Nk_B} - \frac{3}{5}\frac{s_0}{k_B}\right]$$

4.14

$$dH = T\,dS + V\,dP$$

$$\frac{\partial^2 H}{\partial P\,\partial S} = \left(\frac{\partial T}{\partial P}\right)_S = \left(\frac{\partial V}{\partial S}\right)_P$$

4.15 $dH = 0 = T\,dS + V\,dP$ and $T \approx T_0$ and $V \approx V_0$, so $dS = -V_0\,\delta P/T_0$. This is the change of entropy of the Universe.

Chapter 5

5.1 Larger than S_0 because it includes the entropy of the sample.

5.2

$$\text{Probability} \approx \frac{e^{-(E_1 - E_0)/(k_B T)}}{1 + e^{-(E_1 - E_0)/(k_B T)}} = 0.01$$

$T = 2.6 \times 10^4$ K.

5.3 (a) $A = 1/(1 + e^{-\Delta/(k_B T)})$. See Fig. P5.3(a). (b) See Fig 5.3(b).

(c) $\quad U(T) = \Delta e^{-\Delta/(k_B T)}/(1 + e^{-\Delta/(k_B T)})$

See Fig. P5.3(c).

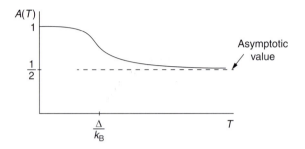

Figure P5.3(a)

$$W_0 = \frac{1}{1 + e^{-\Delta/(k_B T)}} \qquad\qquad W_1 = \frac{e^{-\Delta/(k_B T)}}{1 + e^{-\Delta/(k_B T)}}$$

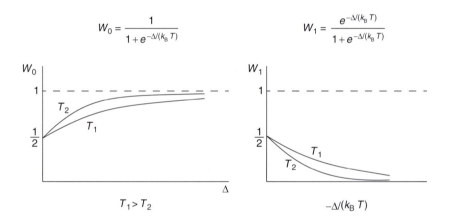

Figure P5.3(b)

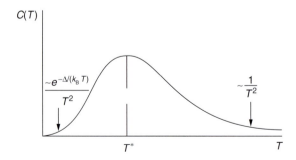

Figure P5.3(c)

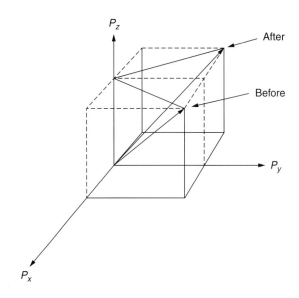

Figure P5.4

(d) $$C = \frac{dU}{dT} = \frac{\Delta^2}{k_B T^2} \frac{e^{-\Delta/(k_B T)}}{(1 - e^{-\Delta/(k_B T)})^2}$$

$$\frac{e^{-x}}{1 + e^{-x}} = \frac{1}{2} - x$$

where $x = k_B T^*/\Delta$; solve for T^*.

5.4 $(p_x, p_y, p_z) \rightarrow (-p_x, p_y, p_z)$. See Fig. P5.4.

5.5

$$\text{Number of states} = \int_0^{k_B T} \rho(\varepsilon) d\varepsilon = \frac{8\sqrt{2}}{3\hbar^2} \pi (kT)^{3/2} V m^{3/2}$$

5.6 $\rho(\varepsilon) = 2\pi m A/h^2$.

5.7 $\rho(\varepsilon) = 4\pi V/(hc)^3$.

5.8 $U = N k_B T$.

5.9 $U = 3 N k_B T$.

5.10 See Fig. P5.10.

5.11

$$S = k_B \log(1 + e^{-\Delta/(k_B T)}) + (\Delta/T)[e^{-\Delta/(k_B T)}/(1 + e^{-\Delta/(k_B T)})$$

As $T \rightarrow \infty$, $S \rightarrow k_B \log 2$.

5.12

$$A = 1 \Big/ \int_0^\infty \rho(\varepsilon) e^{-\varepsilon/(k_B T)} \, d\varepsilon$$

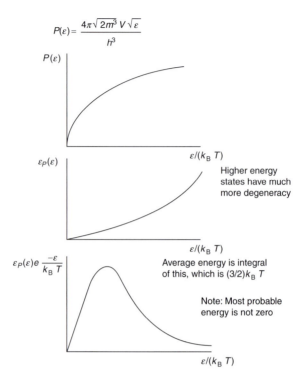

$$P(\varepsilon) = \frac{4\pi\sqrt{2m^3}\,V\sqrt{\varepsilon}}{h^3}$$

Higher energy states have much more degeneracy

Average energy is integral of this, which is $(3/2)k_B T$

Note: Most probable energy is not zero

Figure P5.10

$$U = A \int_0^\infty \varepsilon e^{-\varepsilon/(k_B T)} \rho(\varepsilon)\,d\varepsilon$$

$$S = -kA \int e^{-\varepsilon/(k_B T)} \rho(\varepsilon) \log A\, e^{-\varepsilon/(k_B T)}\,d\varepsilon$$

$$U - TS = k_B T \log A$$

5.13 As $T \to 0$, $S_1(T) \to -\infty$, which is not possible. This fails because of the assumed $k_B T \gg$ energy separation. Instead take ground state plus six degenerate states at energy $\Delta = p_0^2/(2m)$. Then

$$S = -k\left[\left(\frac{1}{1 + 6e^{-\Delta/(k_B T)}}\right) \log\left(\frac{1}{1 + 6e^{-\Delta/(k_B T)}}\right) \right.$$
$$\left. + 6\left(\frac{e^{-\Delta/(k_B T)}}{1 + 6e^{-\Delta/(k_B T)}}\right) \log\left(\frac{e^{-\Delta/(k_B T)}}{1 + 6e^{-\Delta/(k_B T)}}\right) \right]$$

As $T \to 0$

$$S \to (6\Delta/T)e^{-\Delta/(k_B T)} \to 0$$

5.14 You can try this.

5.15

$$S = Nk_B \log\left(\frac{(\pi m k_B T)^{3/2}}{Nh^3}\right)e^{5/2}2\sqrt{2V^2}$$

and $U = (3/2)Nk_BT$.

$$\frac{Nh^2}{2V}\exp\left[\frac{2}{3}\left(\frac{S}{Nk_B} - \frac{5}{2}\right)\right] = \frac{\pi m U}{(3/2)N} \text{ and } s_0/k = 5/3 + \log(2\pi m k_B/h^2)$$

so

$$U = \frac{3}{2}Nk_B\left(\frac{N}{V}\right)^{2/3}\exp\left(\frac{S}{(3/2)Nk_B} - \frac{s_0}{k_B}\right)$$

Chapter 6

6.1

$$J = \frac{V}{C_P}(\beta T - 1) = -\frac{2NV[aN(3bN - 2V)]bPV^2}{5Nk_BPV^3 + aN^3k_B(6Nb - V)}$$

6.2 $P_1 = P_2$ and $\partial P_1/\partial V_1 + \partial P_2/\partial V_2 < 0$, so $K_T > 0$.

6.3 $dU = T\,dS - P\,dV + \mu\,dN$, so U is minimum when S, V and N are constant; $dH = T\,dS + V\,dP + \mu\,dN$, so H is minimum when S, P and N are constant.

6.4

$$F = \sum_i \mu_i N_i - PV$$

$$H = \sum_i \mu_i N_i + TS$$

Chapter 7

7.1 For a single particle,

$$Z_1^{trans} = \left(\frac{2\pi mkT}{h^2}\right)^{3/2}V$$

and $Z = Z_1^N/N!$.

7.2 $Z_1^{vib} = e^{-hv/(2k_BT)}/(1 - e^{-hv/(k_BT)})$.

7.3 $Z_1^{rot} = 2\pi k_B T/\hbar^2$.

7.4 The full partition function is

$$Z = \frac{Z_1^{\text{trans}} Z_1^{\text{vib}} Z_1^{\text{rot}}}{N!}$$

The pressure is still $P = Nk_{\text{B}}T/V$.

7.5 The energy is

$$U = \frac{5}{2}Nk_{\text{B}}T + \frac{1}{2}Nh\nu + \frac{Nh\nu}{e^{h\nu/(k_{\text{B}}T)} - 1}$$

7.6

$$C_V = (\partial U/\partial T)_{N,V} = \frac{5}{2}Nk_{\text{B}} + Nk_{\text{B}}[h\nu/(k_{\text{B}}T)]^2 e^{h\nu/(k_{\text{B}}T)}/(1 - e^{h\nu/(k_{\text{B}}T)})^2$$

7.7

$$\Xi = \sum_{N=0}^{\infty} [e^{\mu N/(k_{\text{B}}T)}(2\pi m k_{\text{B}}T/h^2)^{3/2}V]^N/N!$$

Chapter 8

8.1 We divide by $N!$ only at high enough temperature so that no states are multiply occupied.

8.2

$$Z = \int \exp\left[-\left(\frac{p^2}{2m} + br^{2n}\right)\Big/(k_{\text{B}}T)\right]\frac{d^3r\, d^3p}{h^3}$$

Write

$$\int_0^{\infty} x^m e^{-x}\, dx = \Gamma(m+1)$$

Then

$$Z = \Gamma\left(\frac{3}{2n}\right)\Gamma\left(\frac{1}{2}\right)\frac{(4\pi)^2}{4nh^3}\left(\frac{kT}{b}\right)^{3/(2n)}(2mk_{\text{B}}T)^{3/2}$$

8.3

$$\prod_{i<j}(1 + \varepsilon_{ij}) = 1 + \sum_{i<j}f_{ij} + \sum_{\substack{i<j \\ k<l}}f_{ij}f_{kl} + \cdots$$

$$Q_N^{DG} = V^N + \frac{N(N-1)}{2}V^{N-1}4\pi \int_0^\infty f(r)r^2\,dr$$

$$+ \sum_{\substack{i<j \\ k<l \\ (i,j)\neq(k,l)}} \int d^3x_1 \ldots \int d^3x_n f_{ij}f_{kl} + \cdots$$

$$Q_N^{DG} = V^N + \frac{N(N-1)}{2}V^{N-1}4\pi \int_0^\infty f(r)r^2\,dr + \frac{N^4}{4}V^{N-2}\left[4\pi \int f(r)r^2 dr/2\right]^2$$

$$+\cdots$$

$$Z = Z_{IG}\left[1 - \frac{N^2}{V}B(T) + \frac{N^4}{V^2}B(T)^2 + \cdots\right]$$

$$P = \frac{Nk_BT}{V}\left[1 + \left(\frac{N}{V}\right)B - N\left(\frac{N}{V}\right)^2 B^2 + \cdots\right]$$

8.4 In doing so we are ignoring correlations between the particles.

8.5 The result is given below, as the solution to Problem 8.9.

8.6 The result is given in the main part of the book.

8.7 The result is given in the main part of the book

8.8 The result is given in the main part of the book.

8.9

$$\frac{P_C V_C}{Nk_B T_C} = \frac{[a/(27b^2)]3Nb}{N[8a/(27b)]} = 3/8$$

Index

164